Land Rover Discovery

Owner's Handbook

DISCOVERY
Contents

Published by the
Technical Publications Department
of Land Rover
Publication No. SJR 820 ENHB 90
©Copyright Land Rover 1989

Land Rover
Lode Lane
Solihull
West Midlands, B92 8NW
England

INTRODUCTION

1

MODELS COVERED

The information in this handbook covers all current versions of Land Rover Discovery petrol and diesel models. It is presented in sections to guide the reader progressively from reception of the vehicle through familiarisation with controls and instrumentation, driving techniques and basic day to day attention.

Where specific information is sought, first consult the list of contents (at the front of the book) which will direct you to the relevant page or pages.

Introduction

THE NEW VEHICLE

With every new Discovery special literature is provided which should be read by all owners and drivers to help obtain the best operating results. The literature consists of the following:

1. HANDBOOK: This book, which you are now reading, gives general information about Discovery, also incorporates notes on service, the Vehicle Warranty and full information on how to carry out the necessary day-to-day running maintenance.

2. SERVICE RECORD: Which gives details of the maintenance required and includes spaces for the Dealer to sign and stamp to certify that the work has been carried out at the appropriate intervals.

The operations carried out by your Dealer will be in accordance with current recommendations and may be subject to revision from time to time.

Upon receiving the new Discovery the owner should immediately:

- Examine the Handbook for advice on new features and as an aid to getting the best out of the vehicle.

- Arrange with a Land Rover Dealer to carry out regular maintenance attention.

Introduction

VEHICLE WARRANTY

Land Rover issue under the heading of Vehicle Warranty an undertaking regarding its Service Policy. The Vehicle Warranty is supplied in the Literature Pack and the following notes are given for guidance in the event of a claim being put forward:

1. The Discovery or the part in respect of which a claim is made must be taken immediately to a Land Rover Dealer. This should, wherever possible, be the Dealer responsible for the sale of the vehicle to the owner.

2. The Dealer will examine the parts or vehicle and will without charge advise on the action to be taken in respect of the claim. It will be noted that the Company must reserve the right to examine any alleged defective parts or material should they think fit prior to the settlement of any claim.

3. It must be understood that the factors of wear and tear and any possible lack of maintenance or unapproved alteration will be taken into consideration in respect of any claim submitted.

4. It will be noted that tyres and glass are expressly excluded. The manufacturers of those tyres which the Company fits as standard to its vehicle will always be prepared to consider any genuine claim.

PARTS AND ACCESSORIES

When new parts or accessories are required, obtain Genuine Land Rover parts, or parts supplied through sources approved by the Company.

Land Rover Dealers are obligated to supply only such parts.

Other sources often sell parts suitable for Discovery but frequently these are not made to the same standard or specification as the Company parts and are therefore less likely to give the requisite performance.

Genuine Land Rover parts and accessories are designed and tested for your vehicle and have the full backing of the Land Rover Vehicle Warranty. ONLY WHEN GENUINE LAND ROVER PARTS ARE USED CAN RESPONSIBILITY BE CONSIDERED UNDER THE TERMS OF THE WARRANTY.

Safety features embodied in the vehicle may be impaired if other than genuine parts are fitted. In certain territories, legislation prohibits the fitting of parts not to the vehicle manufacturer's specification. Owners purchasing accessories while travelling abroad should ensure that the accessory and its fitted location on the vehicle conform to mandatory requirements existing in their country of origin.

Introduction

MAINTENANCE ATTENTION

Regular maintenance is one of the main factors in ensuring continuing reliability and efficiency. For this reason detailed schedules have been prepared so that at the appropriate mileages or times owners may know what is required.

The **Pre-delivery Inspection** is a very important first step in the work of preventive maintenance. The Dealer responsible for the sale of the new Discovery will have completed the work involved. There is provision in the Service Record for certification that this work has been carried out.

Normal day-to-day attention required is described in Section 4 of this handbook.

The **first Service Inspection** should be carried out by the Dealer responsible for the sale of the Discovery to the owner at or about 1500 km (1000 miles). A charge is made only for the lubricants etc. used in carrying out the service. Where for any reason it is not convenient for this first service to be carried out by the Dealer responsible for the sale, it can, by prior arrangement, be carried out by any other Land Rover Dealer.

WORKSHOP MAINTENANCE

This is the necessary work to keep the vehicle reliable, and should be done at the following intervals, whichever is first, by a Land Rover Dealer;

200Tdi - 10000 km (6000 miles) or 6 months.

V8   - 20000 km (12000 miles) or 12 months.

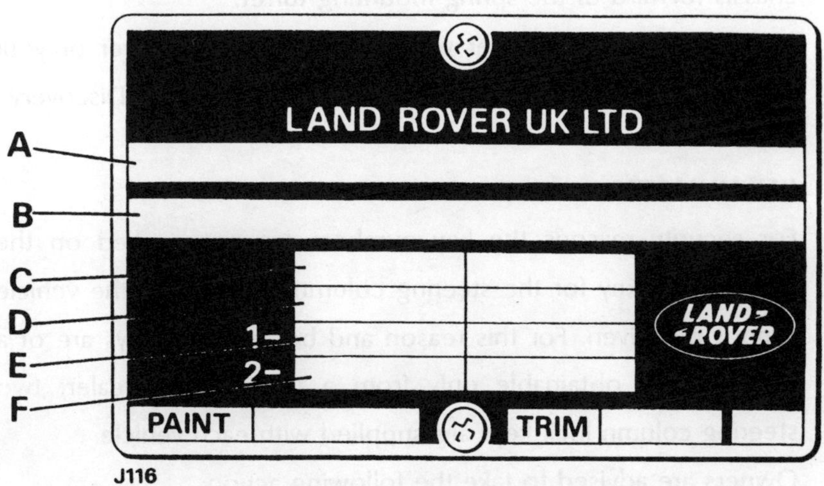

KEY TO VEHICLE IDENTIFICATION NUMBER (V.I.N.) PLATE - Fig. J116

A   Type approval

B   V.I.N. (minimum of 17 digits)

C   Maximum permitted laden weight for vehicle

D   Maximum vehicle and trailer weight

E   Maximum road weight - front axle

F   Maximum road weight - rear axle

Introduction

VEHICLE IDENTIFICATION NUMBER (V.I.N.)

The Vehicle Identification Number and the recommended maximum vehicle weights are stamped on a plate that is rivetted to the top of the radiator grille panel in the engine compartment.

The number is also stamped on the right-hand side of the chassis forward of the spring mounting turret.

Always quote this number when writing to Land Rover or your Distributor and Dealer on any matter concerning your Discovery.

KEY NUMBERS

For security reasons the key numbers are not marked on the locks. If the key for the steering column lock is lost, the vehicle cannot be driven. For this reason and because the keys are of a special type, obtainable only from a Land Rover Dealer, two steering column lock keys are supplied with each vehicle.

Owners are advised to take the following action:

(a) Immediately on receipt of the vehicle, record all the key numbers so that in case of loss, new keys can be obtained.

(b) Keep a spare steering column lock key away from the vehicle in a safe place, but where it is readily accessible.

The steering column lock if properly used reduces the possibility of theft.

FOR YOUR SAFETY

As you read through this manual you will see several places marked **WARNING, CAUTION** or **NOTE** presented in the following form. These have been included to remind you of areas where you should use extra care to avoid personal injury or damage to the vehicle or components.

WARNING: Procedures which must be followed precisely to avoid the possibility of personal injury.

CAUTION: This calls attention to procedures which must be followed to avoid damage to components.

NOTE: This calls attention to methods which make a job easier to perform.

As you attend to your vehicle you may see labels with this WARNING symbol. It means WARNING; DO NOT touch or attempt adjustments until you have read the special instructions contained on the relevent pages in this manual.

A label with this symbol WARNS that there are very high voltages in the ignition system on some models. DO NOT touch any ignition component while the ignition is switched on, especially where you see this label.

Introduction

WARNING: Your Discovery has a higher ground clearance and hence a higher centre of gravity than an ordinary passenger car to enable it to perform in a wide variety of off-road applications. An abrupt manoeuvre at an inappropriate speed or on an unsuitable surface could cause the vehicle to go out of control.

WARNING: DO NOT remove the expansion tank filler cap when the engine is hot, because the cooling system is pressurised and personal scalding could result.

WARNING: Many liquids and other substances used in motor vehicles are poisonous, they must not be consumed under any circumstances and must be kept away from open wounds. These substances include anti-freeze, brake fluid, fuel, windscreen washer additives, lubricants, battery contents and various adhesives.

WARNING: Keep hands, hair and clothing well clear of the fan blades and other rotating parts, when the engine is running, to avoid the possibility of personal injury.

WARNING: DO NOT mix Cross-ply and Radial-ply tyres on this vehicle. Recommended tyre replacements are given in the General Data section.

Introduction

⚠️ **WARNING: Additions, alterations or repairs to the electrical or fuel systems can create fire hazards if carried out incorrectly. Adhere strictly to methods described in this Owner's Manual or to instructions supplied with genuine Land Rover parts.**

SAFETY

In the interests of safety, your attention is drawn to the following safety hints.

- Regular servicing, including day to day servicing by the driver/owner as described in section 4. and the longer term workshop maintenance, which should be carried out by a Land Rover dealer, is essential to help provide safe, dependable and economic motoring, and to ensure that the vehicle conforms to the various safety regulations in force.

- Always use the seat belts, even for the shortest journeys.

- Before driving, learn the layout and use of all controls, gears and switches.

- Before driving the vehicle adjust the seat as necessary to achieve a comfortable driving position with full control over the vehicle.

- Always start vehicle and operate controls from the driving position.

Introduction

- Ensure that the vehicle speed is low enough for an emergency stop to be made safely under all road and vehicle loading conditions.

- Never leave unsupervised children in the vehicle.

- Keep the windscreen, rear and side windows clean to give clear vision. Use a solvent in the screen washer reservoir.

- Maintain all external lights in good working order and correct setting of headlamp beams.

- **DO NOT** turn the ignition/steering lock key to the lock position or try to remove the key whilst the vehicle is in motion.

- Maintain correct tyre pressures. These should be checked at least each month, or more often when high-speed touring or under cross-country conditions, even to the extent of a daily check.

⚠️ **WARNING: Some components on your Discovery, such as gaskets and friction surfaces (brake linings and clutch discs) may contain asbestos. Breathing asbestos dust is dangerous to your health. You are therefore advised to have any maintenance or repair operations on such components carried out by a recognised Land Rover dealer or distributor. If, however service operations are to be undertaken on parts containing asbestos, the following essential precautions must be observed.**

- Work out of doors or in a well ventilated area wear an approved protective breathing mask.

- Dust found on the vehicle or produced during work on the vehicle should be removed by extraction and not by blowing.

- Dust waste should be dampened, placed in a sealed container and marked to ensure safe disposal.

- If any cutting, drilling etc., is attempted on materials containing asbestos the item should be dampened and only hand tools or low speed power tools used.

For your further guidance, Land Rover replacement parts which contain asbestos are progressively being identified by the symbol shown below. If you are in any doubt, please consult your dealer or distributor.

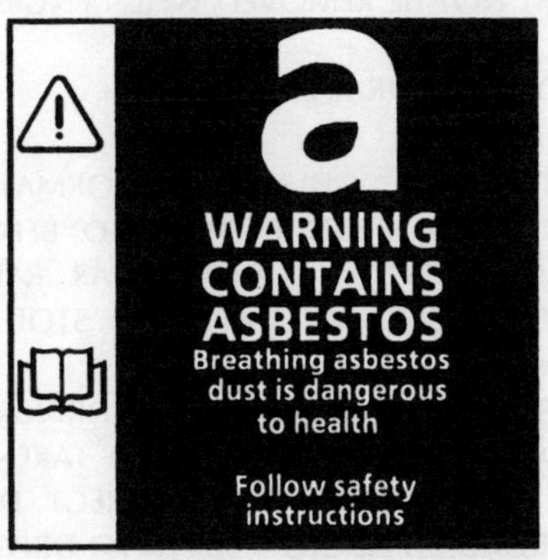

Introduction

LABELS FIXED TO THE VEHICLE - Fig. J112 and J113

Various labels are fixed to the vehicle to draw the drivers and owners attention to important and helpful information. The following illustration shows the location of the labels together with a translation. Check the position of the labels and the translated contents.

1.   LAND ROVER
IMPORTANT - BEFORE JACKING VEHICLE
1. ENGAGE DIFF. LOCK (i.e. WARNING LIGHT MUST BE ILLUMINATED PRIOR TO SWITCHING OFF IGNITION)
2. APPLY HANDBRAKE
3. CHOCK WHEELS

2.   WARNING CONTAINS ASBESTOS - Breathing asbestos dust is dangerous to health, follow safety instructions

3.   HEADLAMP LEVELLING (WHEN FITTED)

4.   THIS PLUG MUST NOT BE REMOVED WHEN ENGINE IS HOT

5.   ANTI FREEZE - DO NOT DRAIN

6.   IMPORTANT TRANSFER GEARBOX INFORMATION. TO CHANGE TRANSFER BOX RATIO, REDUCE SPEED TO BELOW 8 KPH (5 MPH), SELECT NEUTRAL, MOVE HIGH/LOW GEAR RAPIDLY TO REQUIRED POSITION, SELECT GEAR. ALTERNATIVELY, STOP VEHICLE AND MAKE SELECTION AS ABOVE.
WARNING THIS VEHICLE MUST NOT BE DRIVEN ON A TWO WHEEL ROLLING ROAD UNLESS PRECAUTIONS ARE TAKEN.
FOR FURTHER INFORMATION ON CORRECT DRIVING TECHNIQUES AND ROLLING ROAD OPERATION, REFER TO DRIVER'S MANUAL.

Introduction

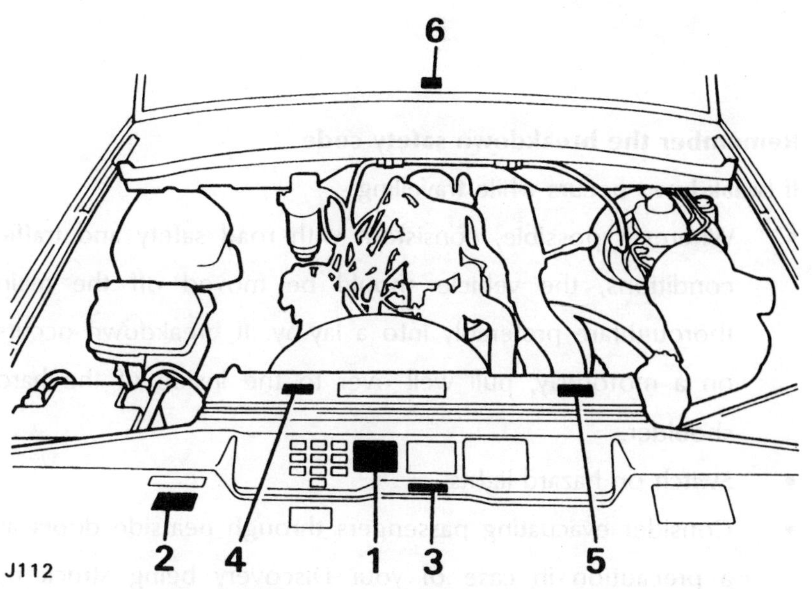

Location of labels - 4 cylinder 200Tdi - left hand steering

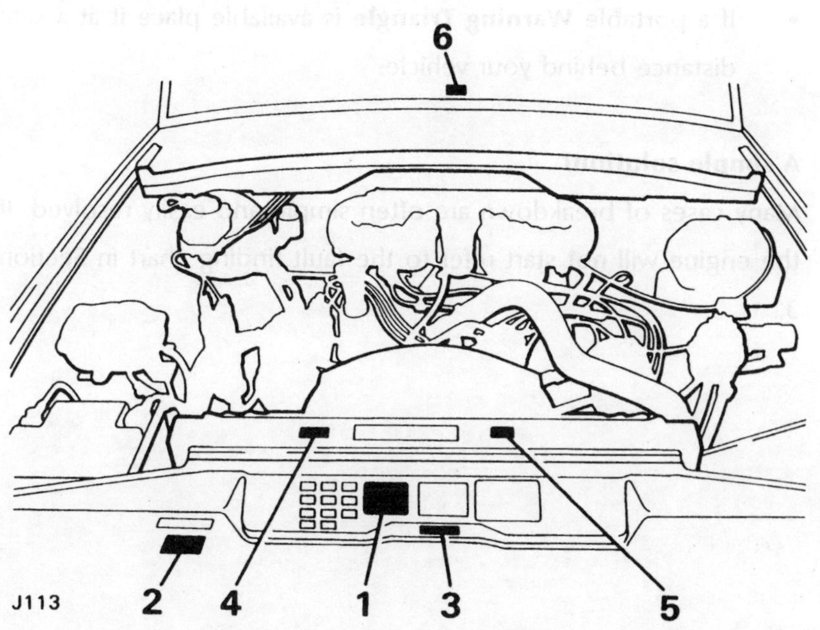

Location of labels - V8 petrol - right hand steering

Introduction

Remember the breakdown safety code

If breakdown occurs while travelling:-

- Wherever possible, consistent with road safety and traffic conditions, the vehicle should be moved off the main thoroughfare preferably into a lay-by. If breakdown occurs on a motorway, pull well over to the inside of the hard shoulder.

- Switch on hazard lights.

- Consider evacuating passengers through nearside doors as a precaution in case of your Discovery being struck by another vehicle.

- If a portable **Warning Triangle** is available place it at a safe distance behind your vehicle.

A simple solution?

Many cases of breakdown are often simple and easily resolved. If the engine will not start refer to the fault finding chart in Section 3.

CONTROLS AND INSTRUMENTS

2

Controls and instruments

INTRODUCTION

General

This section helps to explain the controls and instruments on your vehicle.

"One picture is worth a thousand words" as the saying goes. We at Land Rover have attempted to put this into practice, therefore this section is mainly self explanatory with illustrations and as few a words as possible.

Before driving your vehicle, it is important for your safety that you read and thoroughly understand how to use all the controls.

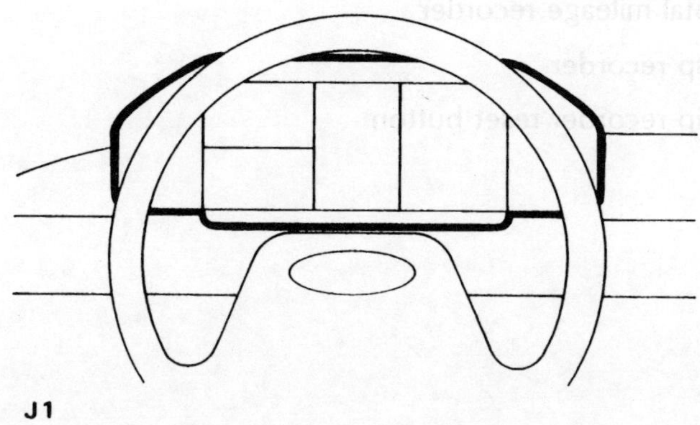

Binnacle location

Speedometer

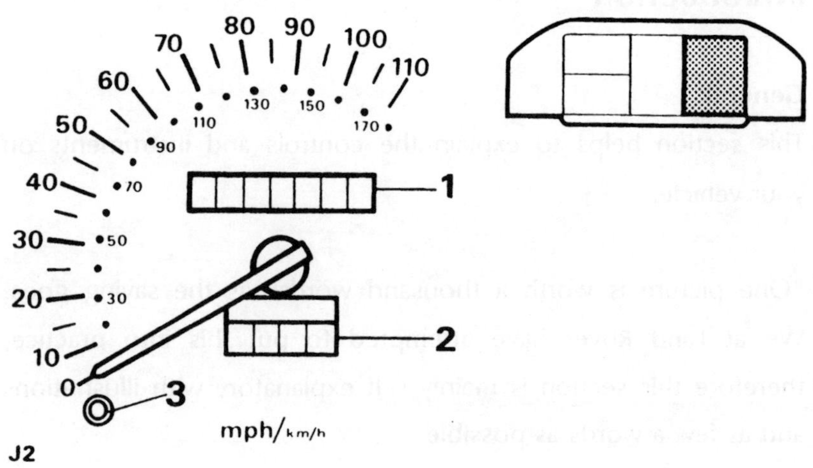

SPEEDOMETER - Fig. J2

1.   Total mileage recorder

2.   Trip recorder

3.   Trip recorder reset button

Revolutions counter

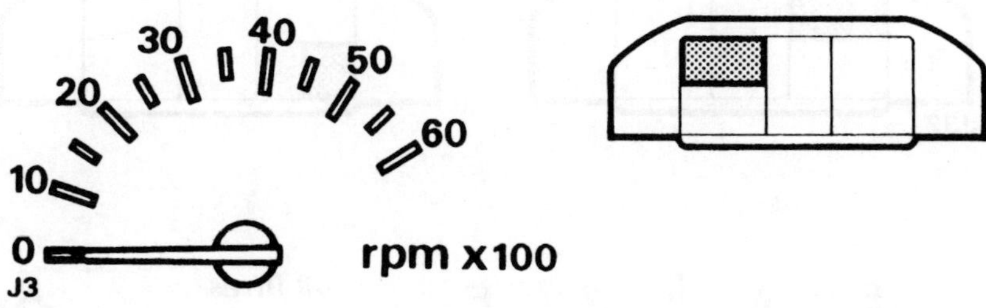

ENGINE REVOLUTIONS COUNTER (when fitted) - Fig. J3

Under normal conditions the engine is most fuel efficient within the range of approximately 2000 to 3000 r.p.m. Higher speeds can be used briefly, while accelerating before changing up.

Fuel gauge

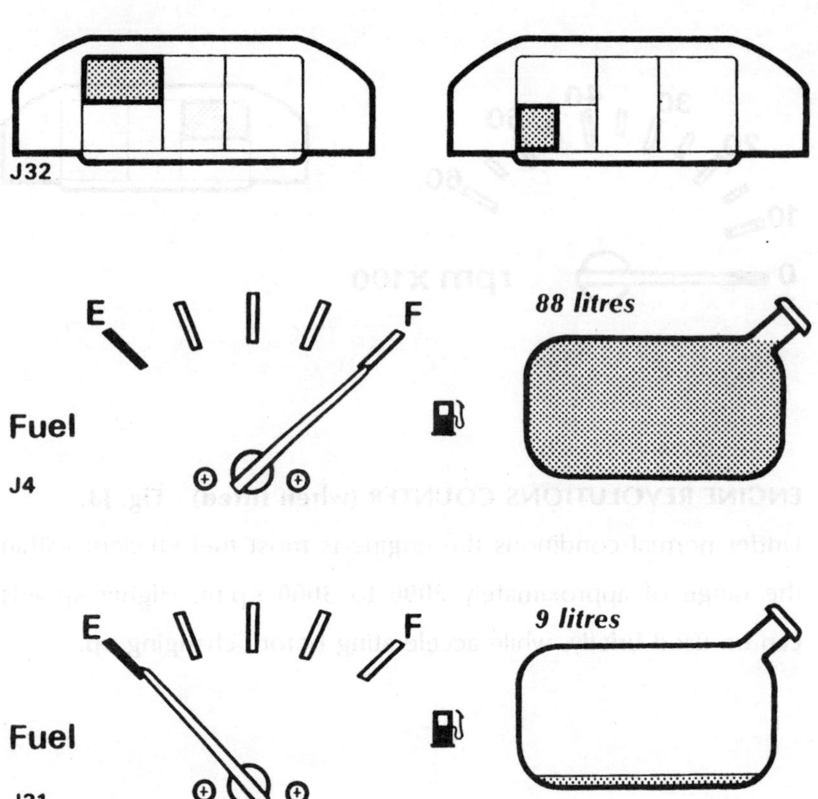

FUEL GAUGE - Fig. J32, J4 and J31

The fuel gauge shows the approximate contents of the tank. The style and position of the gauge within the binnacle may vary from one model to another as shown.

FUEL WARNING LIGHT (when fitted) Fig. J6

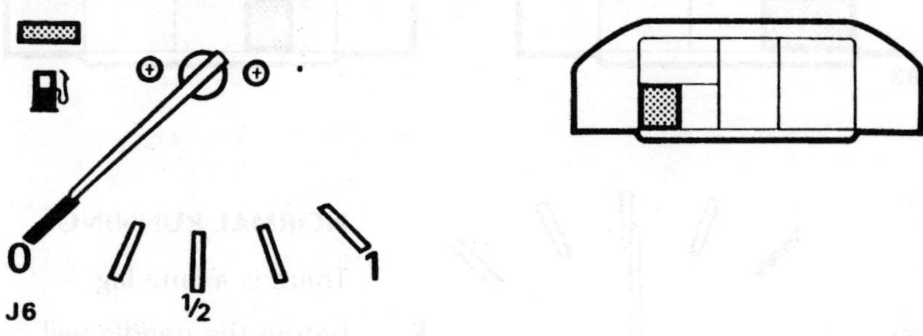

The amber light will illuminate to remind you that the fuel tank needs refilling. There is approximately 9 litres left in the tank.

NOTE: Turn starter key to position "O" to put out the warning light before refueling.

Coolant temperature

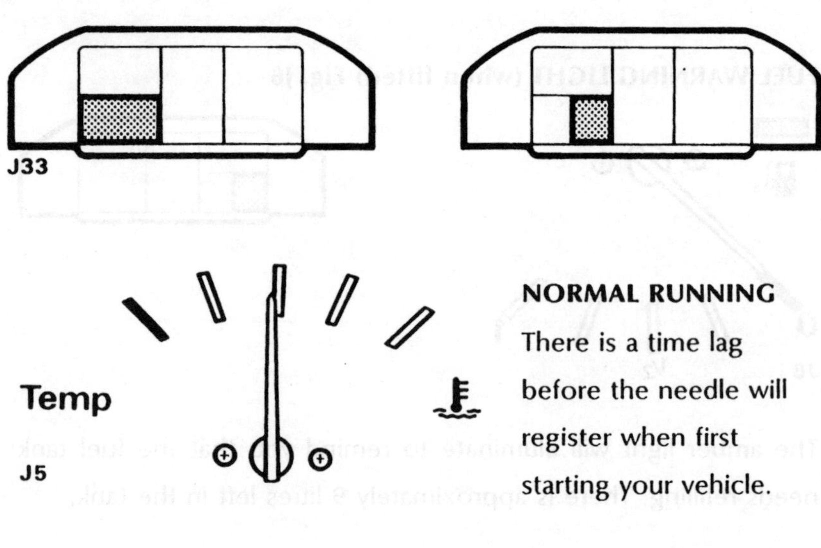

NORMAL RUNNING

There is a time lag before the needle will register when first starting your vehicle.

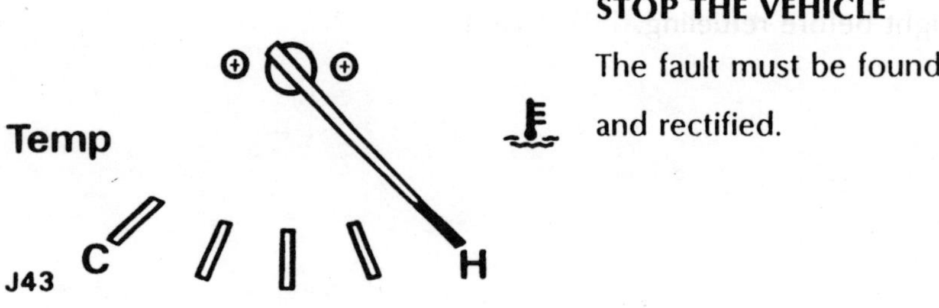

STOP THE VEHICLE

The fault must be found and rectified.

WARNING: DO NOT remove the expansion tank filler cap when the engine is hot, because the cooling system is pressurised and personal scalding could result.

The style and position of the gauge in the binnacle will vary from one model to another as shown.

Coolant temperature warning light

COOLANT TEMPERATURE WARNING LIGHT (when fitted)

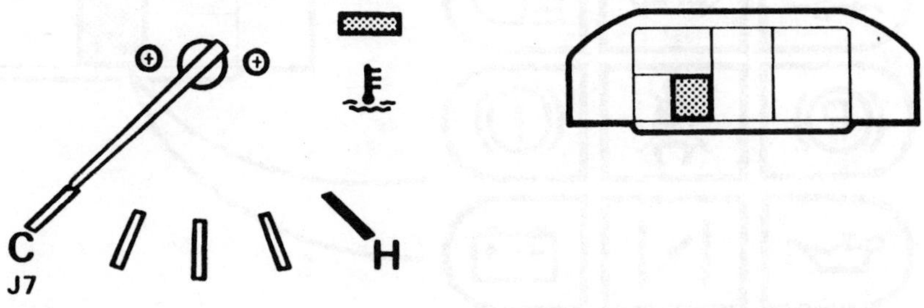

If the red light flashes, the engine may be overheating. The vehicle should be stopped immediately and the cause investigated and repaired.

⚠ **WARNING: DO NOT remove the expansion tank filler cap when the engine is hot, because the cooling system is pressurised and personal scalding could result.**

Warning lights panel

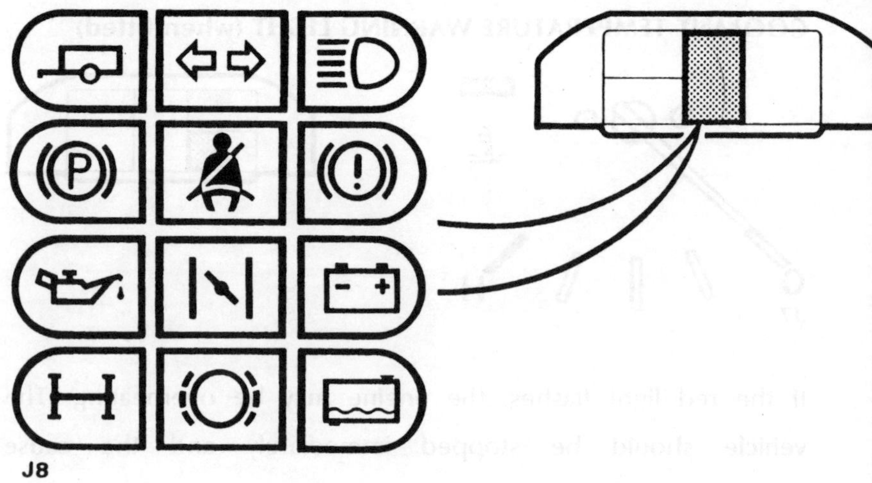

J8

WARNING LIGHTS PANEL - Fig. J8

There are twelve warning lights using four different colours.

RED - WARNING

DO NOT drive the vehicle if a red warning light is continuously on, while the engine is running. The cause must be found and rectified. If in doubt, seek expert advice.

ORANGE - CAUTION

Shows unit is operating. Should be switched off (or rectified) as soon as conditions allow.

GREEN AND BLUE

Shows unit is operating.

TRAILER WARNING LIGHT - Fig. J9

J9

The green warning light will come on when a trailer plug is connected to the vehicle trailer socket. It will flash in conjunction with the vehicle indicator warning lamps, showing that the trailer indicator lamps are functioning correctly. In the event of a bulb failure on the trailer, the warning light will flash once only and then remain off. Change the bulb as soon as possible. Where a trailer is not connected, the trailer warning light momentarily flashes when the direction indicator switch is operated.

Direction and main beam

DIRECTION INDICATOR LIGHT - Fig. J12

J10

The green warning light will come on when the direction indicators are working. If the warning light does not operate, there may be a bulb failure in the warning light panel or in one of the lights. Have the bulb changed as soon as possible.

MAIN BEAM WARNING LIGHT - Fig. J11

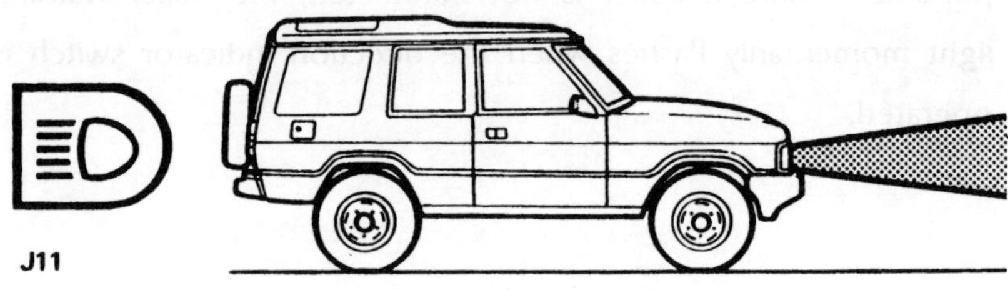

J11

The blue warning light will come on when the headlamp main beams are in use. The warning light will also come on when flashing the headlamps.

Park brake and seat belt

PARK BRAKE WARNING LIGHT - Fig. J12

J12

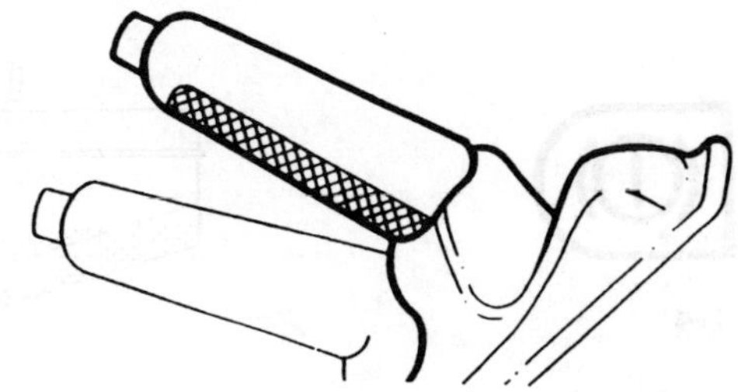

The red warning light will come on when the handbrake lever is applied with the engine starter key in position "II". If this does not happen replace the bulb.

SEAT BELT WARNING LIGHT (when fitted)

The red warning light will come on and remain on when the drivers or passengers seat belts are not fastened with the engine starter key in position "II". The driver and passengers should always fasten their safety belts before driving away.

Brake fluid pressure

BRAKE FLUID PRESSURE FAILURE WARNING LIGHT - Fig. J14

J14

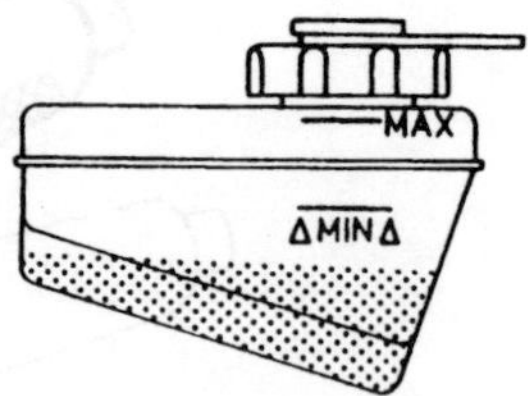

The red warning light should come on as a bulb check when the handbrake lever is applied with the engine starter key in position "II". If this does not happen replace the bulb (Section 4). It will also come on when the fluid drops below the minimum permissible level. If the warning light does not go out or comes on during normal running, the vehicle should be stopped immediately, the cause investigated and repaired. As a fail-safe precaution against total brake failure, a primary and secondary hydraulic system is incorporated. Should one of the hydraulic circuits fail the other circuit will continue to function. This will result in increased brake pedal travel and effort, also longer stopping distances.

Brake fluid pressure

 **WARNING: DO NOT pump the brake pedal in an attempt to restore pedal pressure.**

If there is pressure failure in one of the brake circuits the cause must be investigated immediately.

You should leave the vehicle where it is and call for assistance. Even if you are certain that it is safe to proceed, extreme care should be taken and heavy braking avoided. In deciding whether it is safe to proceed you must consider whether you will be infringing the law.

Oil pressure

OIL PRESSURE WARNING LIGHT

The red warning light will come on as a bulb check when the starter key is turned to position "II" and go off when the engine is running above idling speed. The light may flicker at idling speed, but if it remains on during normal running, the vehicle should be stopped immediately and the cause investigated and rectified.

COLD START WARNING LIGHT (200Tdi) - Fig. J16

J16

The amber warning light will come on when the engine starter key is turned to the position "II" on 200Tdi models. The light will go off as soon as the correct starting temperature has been reached.

COLD START WARNING LIGHT (petrol) - Fig. J17

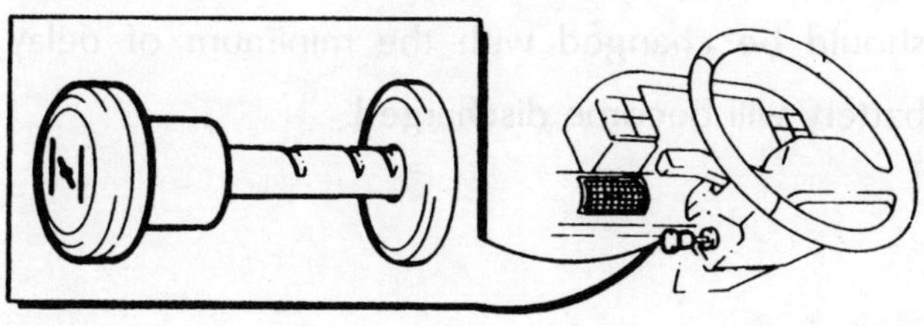

J17

The amber warning light will come on when the cold start control is pulled out with the engine starter key in position "II". The light will go off when the control is pushed in, which should be done as soon as conditions permit.

Charge

CHARGE WARNING LIGHT

The red warning light will come on as a bulb check when the starter switch is turned to position "II".

NOTE: If the warning light does not go out, or comes on during normal running, the vehicle should be stopped immediately, the cause investigated and repaired. The charge warning light is connected in series with the alternator field circuit. Bulb failure would prevent the alternator charging, therefore the bulb should be checked before suspecting an alternator fault. A failed bulb should be changed with the minimum of delay otherwise the battery will become discharged.

DIFFERENTIAL LOCK WARNING LIGHT - Fig. J19

J19

The amber warning light will come on when the gearbox differential lock is engaged.

CAUTION: If the warning light remains on with the diff lock disengaged the transmission is 'wound up'. The vehicle must be stopped and reversed a short distance to 'unwind' the transmission. The warning light will then be extinguished and the vehicle can proceed as normal. If, after reversing, the light remains on, consult your dealer as soon as possible.

Brake pad wear

BRAKE PAD WEAR WARNING LIGHT

The amber warning light will come on as a bulb check when the handbrake lever is applied with the starter key is turned to position "II". If this does not happen replace the bulb (Section 4). It will also come on when any of the brake pads have worn down to the minimum permissible thickness. The pad or pads must be renewed as quickly as possible.

Heated rear window, Rear fog and Rear window wipe

These switches are of a push/push type; to operate, push in to switch on and push again to switch off.

Rear window wash

This switch is of a pressure type; push in to switch on and release to switch off.

Radio (when fitted)

Radio volume up, Radio volume down, Seek/search and Waveband select (When fitted). These switches are of a pressure type; push in to operate and release to stop. For further information see the section on Radios.

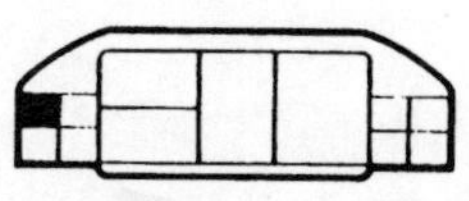

Heated rear window - Fig. J136

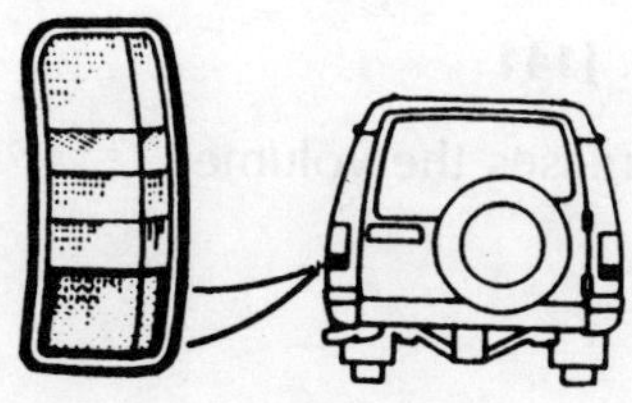

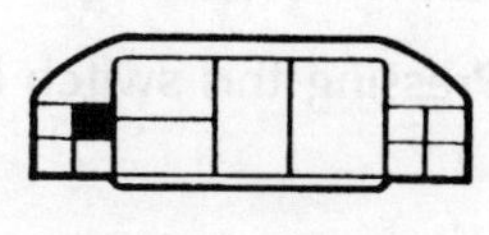

Rear fog guard lamps - Fig. J137

Panel switches

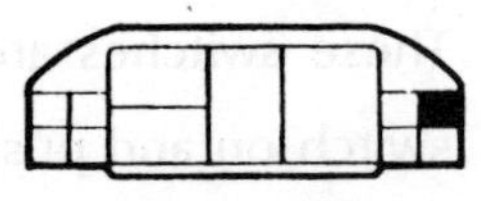

Rear window wipe - Fig. J138

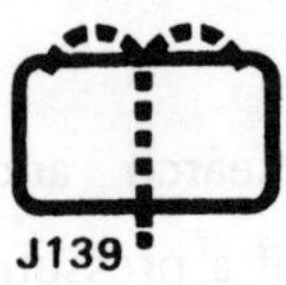

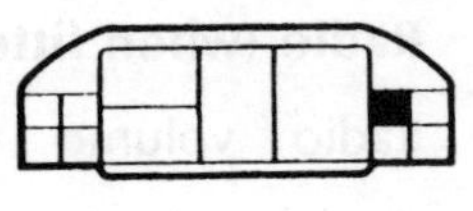

Rear window wash - Fig. J139

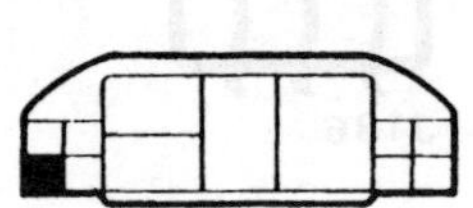

Radio volume up - Fig. J141

Pressing this switch increases the volume.

 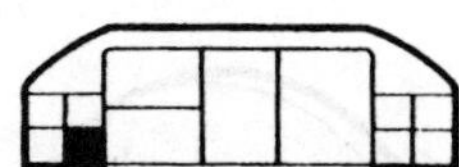

Radio volume down - Fig. J140

Pressing this switch decreases the volume.

 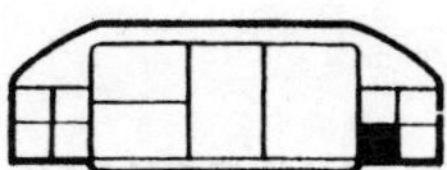

Seek search - Fig. J142

Pressing this switch searches for the next radio station on the waveband (see radio booklet).

 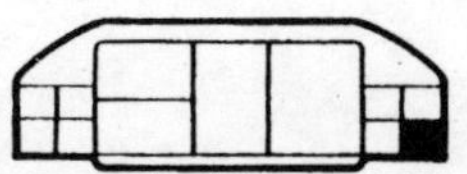

Wave band - Fig. J143

Pressing this switch changes the waveband (see radio booklet).

Steering lock

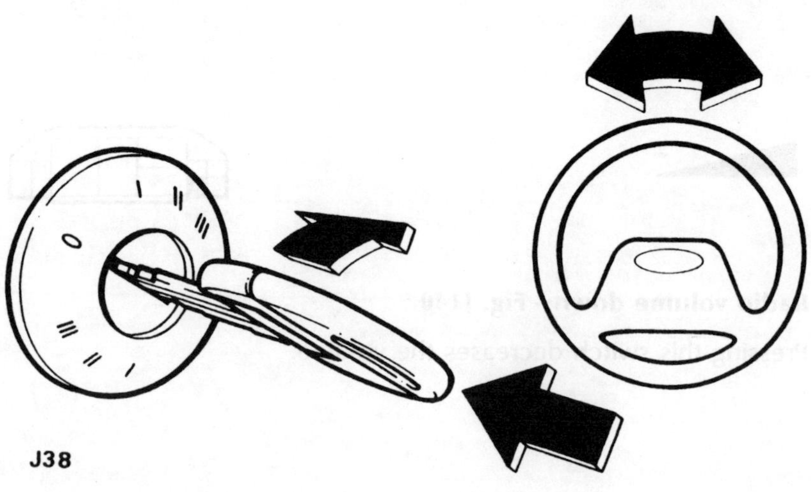

J38

TO UNLOCK THE STEERING COLUMN - Fig. J38

To unlock the steering, insert the key and turn to the first position while moving the steering wheel slightly. This will assist you in its disengagement.

J37

TO LOCK THE STEERING COLUMN - Fig. J37

To lock the steering, turn the key fully back and withdraw it from the lock. Movement of the wheels towards the straight ahead position will engage the steering lock.

WARNING: To prevent the steering column lock engaging it is most important that before the vehicle is moved in any way, for example, being towed or coasting, the key MUST be inserted in the lock and turned to the first position. If, due to an accident or electrical fault it is not considered safe to turn the key, the battery must first be disconnected, then turn the key. DO NOT attempt to remove the key while the vehicle is in motion.

Engine starter switch

ENGINE STARTER SWITCH

The engine starter switch is located to the right of the steering column and is key operated. The switch has four positions and are as follows:

- Steering locked (when key has been taken out)
- Sidelamps operable
- Headlamps: main beam, flash and dipped operable.
- Rear fog guard lamps operable with headlamps on main beam or dipped only.

- Instrument lights operable with sidelamps or headlamps on only.
- Hazard warning lamps operable
- Horn operable.

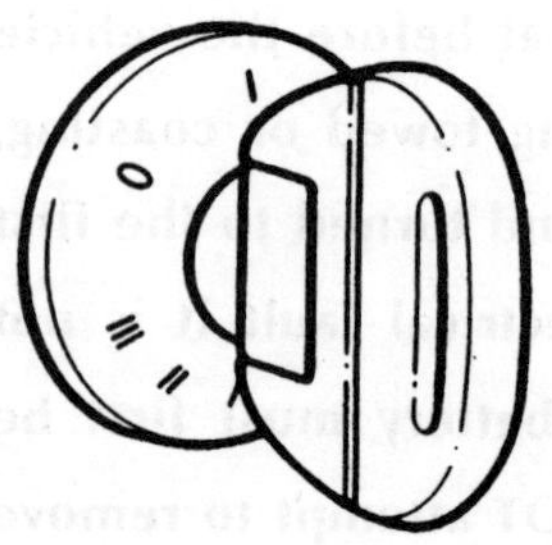

- Steering unlocked.
- Radio/cassette operable.
- Electric windows (when fitted) operable.
- Blower motor fan operable.
- Front wash/wipe switches operable.

- Rear wash and wipe switches operable.

Engine starter switch

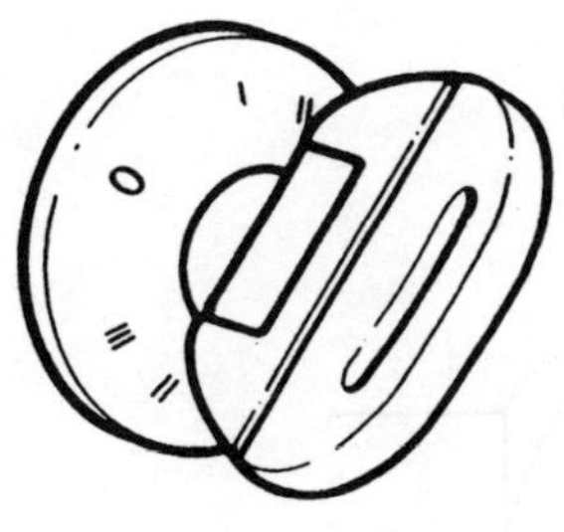

- Cigar lighter.
- Clock.
- Electric mirrors (when fitted).
- Heated rear window.
- Auxiliary instrument lights operable. The warning lights will illuminate when necessary.

The oil pressure and charge warning lights will come on as a bulb check and go off when the engine is running.

NOTE: On 200Tdi engines the cold start warning light will extinguish when the correct starting temperature has been reached. On petrol engines the cold start warning light will come on only if the cold start control is pulled out.

- Starter motor operates.

Release the key immediately the engine starts, the key will automatically return to position "II".

Wash/wipe switches

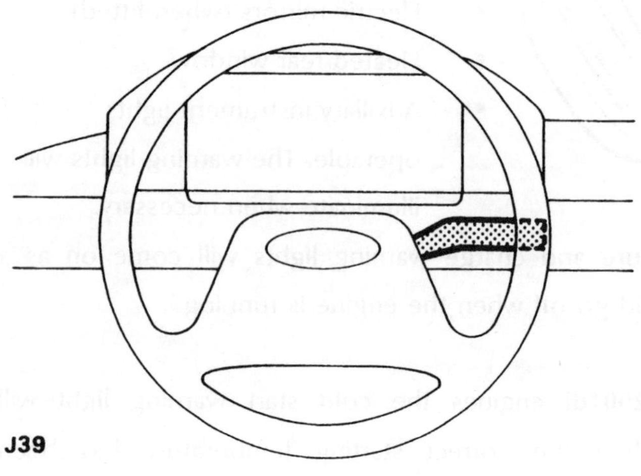

J39

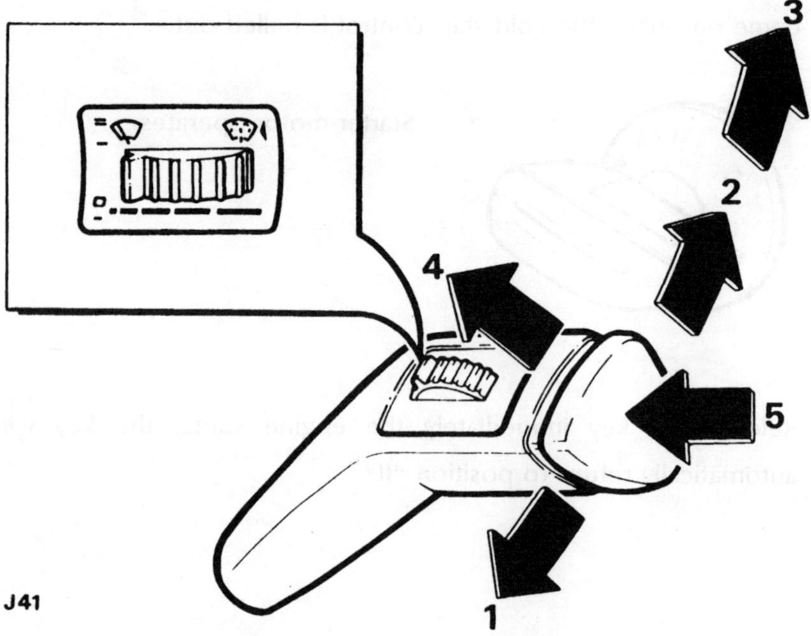

J41

Wash/wipe switches

Front wash/wipe switches

1. When you pull the switch towards you, the wiper blades will travel once across the windscreen.

2. When you move the switch downwards, the wiper blades will travel across the windscreen at regular intervals with a pause between each wipe. The pause between wipes of the screen can be varied using the regulator in the middle of the switch.

3. When you move the switch upwards to the first position, the wiper blades will travel across the windscreen at normal speed.

4. When you move the switch upwards again to the second position, the wiper blades will travel across the windscreen at a faster speed.

5. Push the end of the switch inwards and the washers are operated in this position. The wipers make two or three sweeps after the switch is released.

Lighting switches

LIGHTING STALK - Fig. J40 and J42

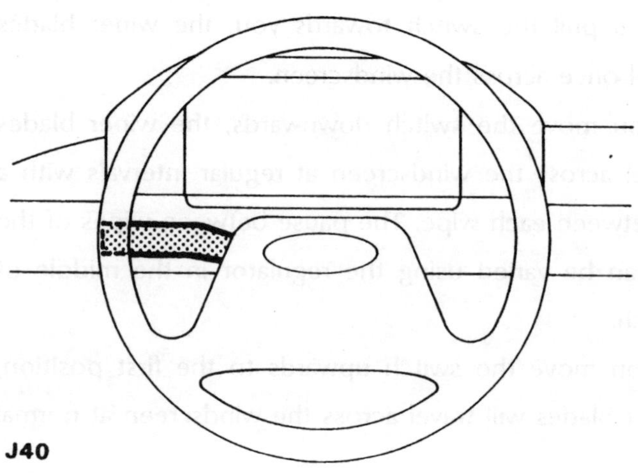

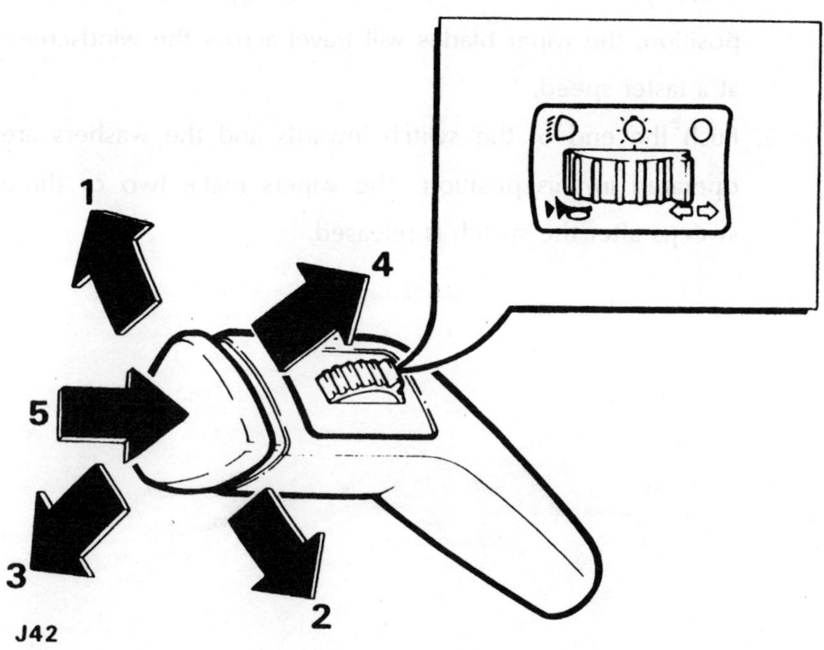

Lighting switches

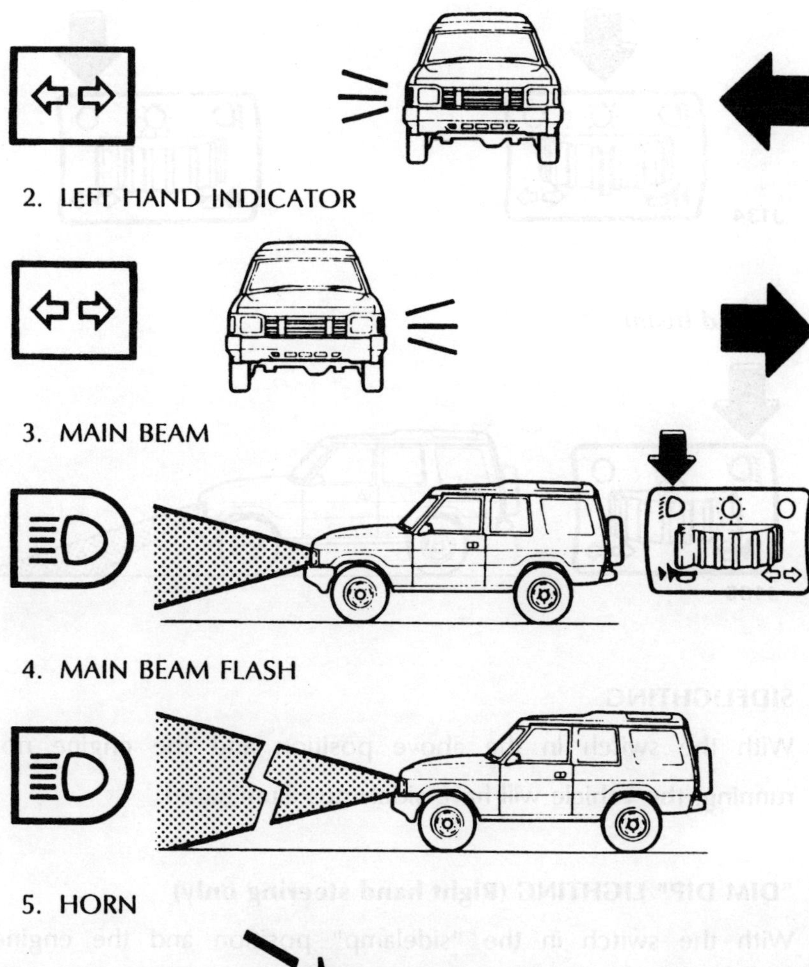

Lighting

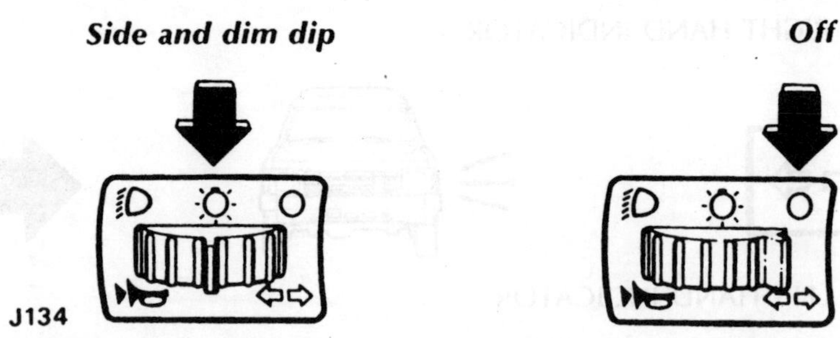

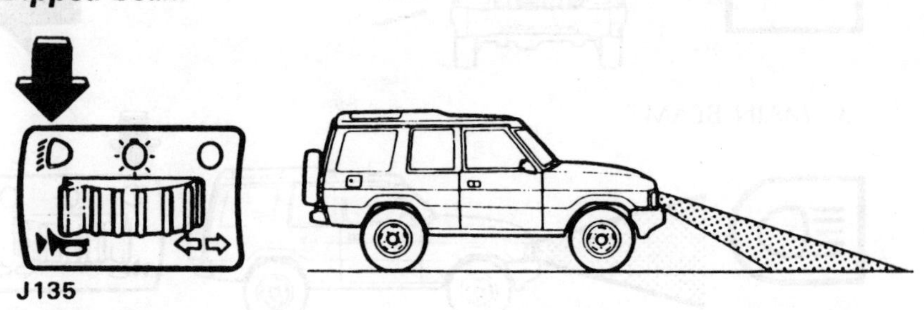

SIDELIGHTING

With the switch in the above position and the engine not running, the vehicle will have sidelamps illuminated.

"DIM DIP" LIGHTING (Right hand steering only)

With the switch in the "sidelamp" position and the engine running, a low current is also supplied to the dipped beam circuit giving dim dipped lighting in addition to sidelamps.

Headlamp levelling

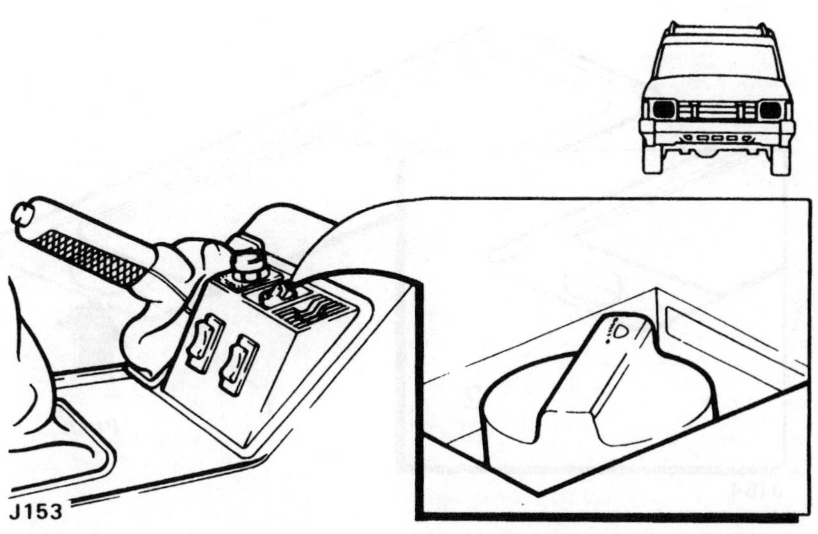

HEADLAMP LEVELLING (German market only) - Fig. J153

The headlamp levelling switch should be turned to one of the four positions depending on the load being carried. The following switch positions should be used as a general guide.

O   Driver only or driver and front passenger only (loadspace empty).

I   Driver, front passenger and two rear passengers (load space empty).

II   All seats occupied by adults and loaded to gross vehicle weight.

III   Driver only plus weight in load space up to maximum rear axle weight (see Section 5).

Interior lights

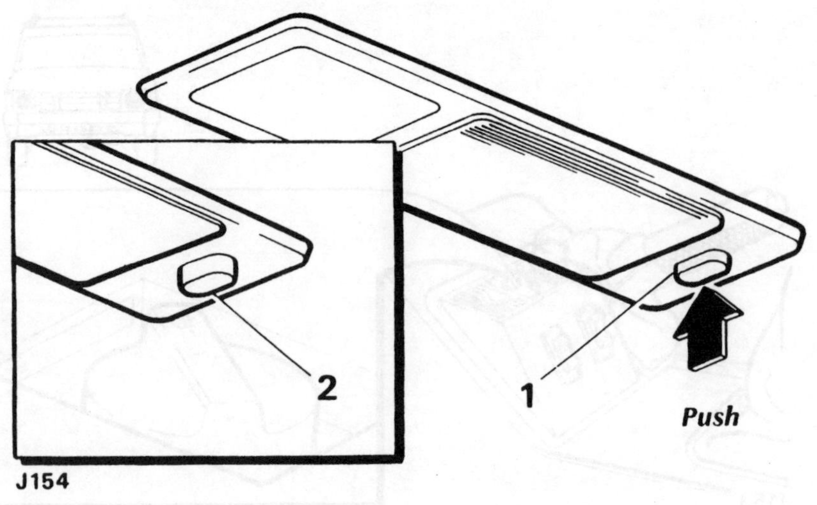

INTERIOR LIGHTS - Fig. J154

Both interior lights will come on when the front doors are opened with the switch in position 1. Only the rear light will come on when the rear door is opened. With the switch in position 2, the light will remain on.

HEADLAMP POWER WASH - Fig. J155

The headlamp power wash will operate, with the headlamps in the dipped beam position, when the windscreen wash switch is pressed. It will operate for one second. Continual pressing of the windscreen wash switch will result in continual headlamp wash.

The jet direction can be altered by inserting a needle into the orifice (1). If they become blocked a needle can be used to unblock them.

Heater and air conditioning

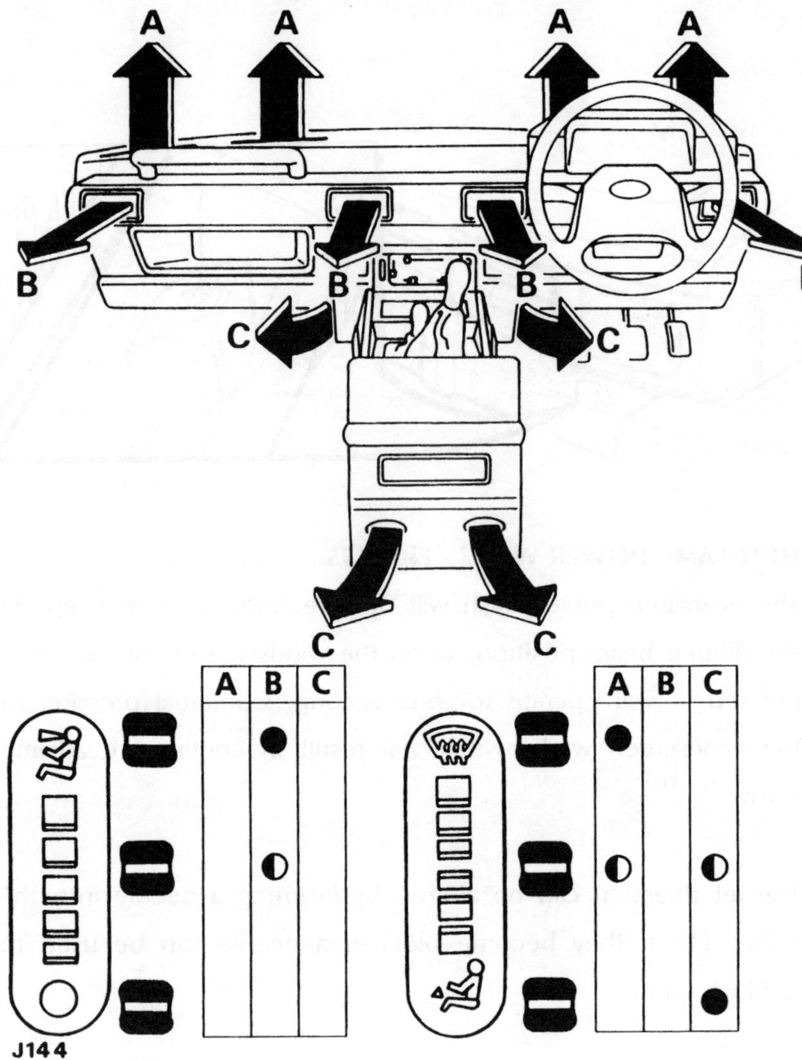

HEATER AND AIR CONDITIONING (where fitted) - Fig. J144

Moving the sliders to the above positions will provide air flow to
A, B or C.

●   All air flow

◐   Partial air flow

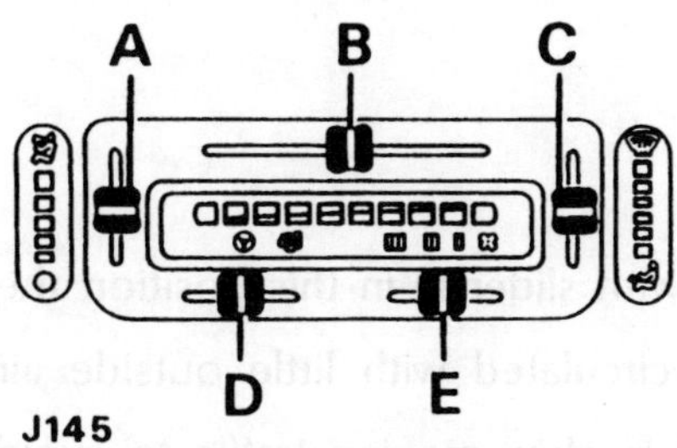

HEATER CONTROLS - Fig. J145

A. Air supply to fascia mounted vents

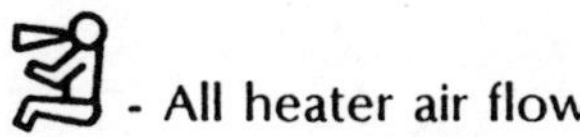 - All heater air flow

○ - No heater air flow

B. Temperature selector

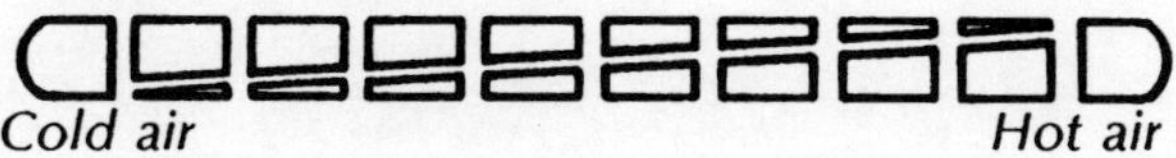

C. Air supply to windscreen and footwells

 Windscreen

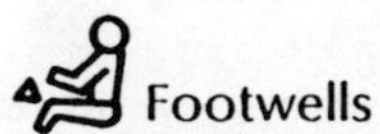 Footwells

Heater controls

D. Air supply selector

Internally recirculated air - with slider D in this position the air inside the vehicle is recirculated with little outside air entering. This can be used in slow moving traffic to avoid fumes.

Fresh outside air - with slider D in this position air is taken into the vehicle from outside.

E. Air blower speed selector

The blower has three speeds that can be selected.

Air conditioning controls

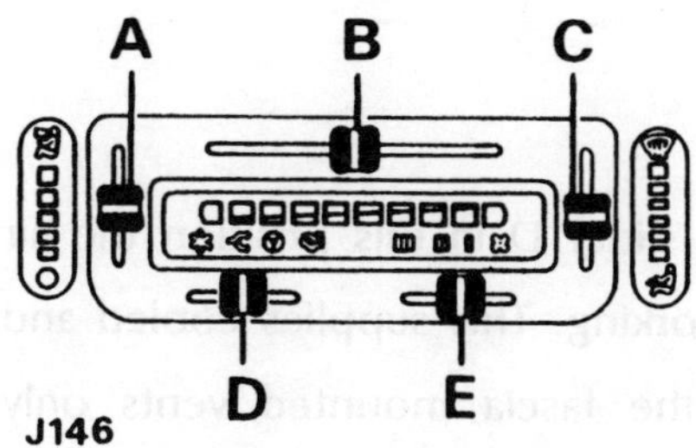

AIR CONDITIONING CONTROLS - Fig. J146

A.   Air supply to fascia mounted vents

Air conditioning is most effective with the control in this position. It stops heater air flowing and allows cooled and dried, recirculated air to pass through the fascia mounted vents.

B.   Temperature selector (see heater).

C.   Air supply to windscreen and footwells (see heater).

Air conditioning controls

D.   Air supply selector

Air conditioning - with the slider D in this position the air conditioning unit will be working. This supplies cooled and dried, recirculated air to the fascia mounted vents only when a blower speed is selected.

Air conditioning combined with heat facility - with the slider in this position it is possible for the air conditioning and the heater to work together. Cooled and dried, recirculated air is supplied to the fascia mounted vents and heated air is supplied to the windscreen or footwells.

Recirculated air (see heater).

Fresh outside air (see heater)

E.   Blower speed selector (see heater)

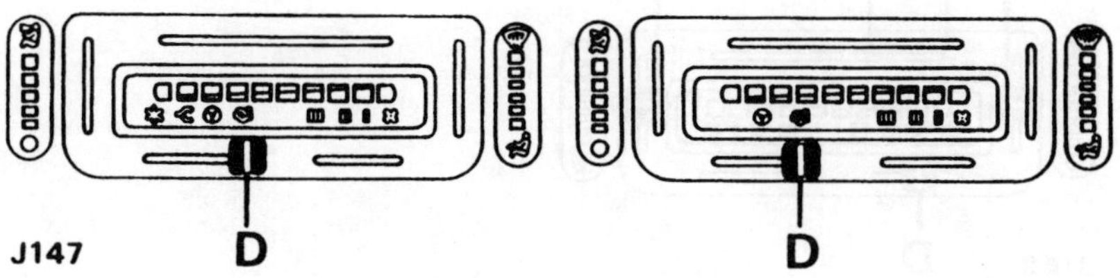

HEATING THE INTERIOR - Fig. J147

Slider D set to fresh outside air position. All other sliders are set to suit temperature and where heater air is needed.

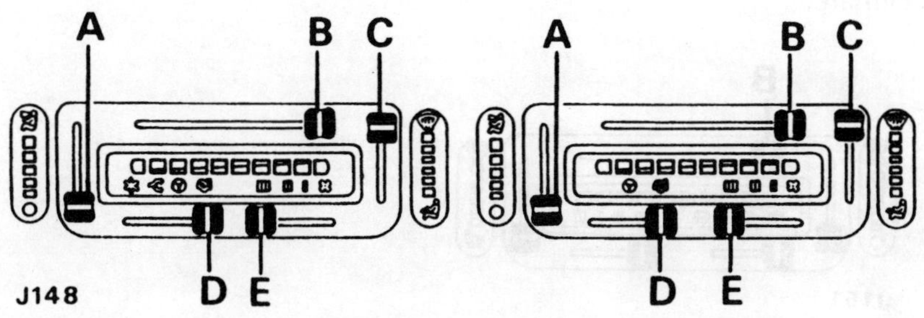

DEMISTING AND DEFROSTING - Fig. J148

With the sliders in these positions demisting the interior and defrosting the exterior are most effective when the engine has warmed up.

Using the air conditioning

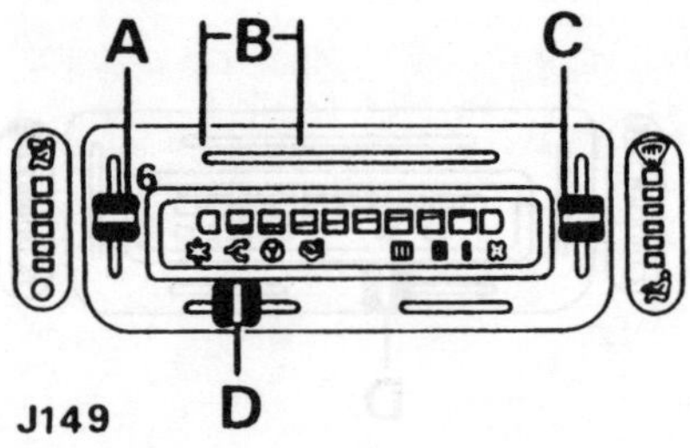

NORMAL COOLING - Fig. J149

Sliders A, C and D should be set as shown. Slider B should be set within the area shown and the blower speed set to suit the climate.

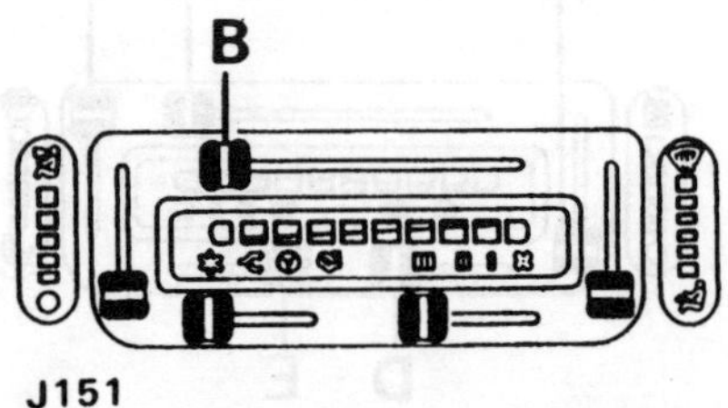

MAXIMUM COOLING - Fig. J151

Set all sliders to the above positions and open a window. After a few minutes the air should be cooler. The window should then be shut and the blower speed adjusted. When the vehicle is at the required temperature, move slider B slightly to the right.

Using the air conditioning

CAUTION: DO NOT use the air conditioning for long periods with the windows or sunroofs (when fitted) open. The system would be working ineffectively at maximum output which could result in component damage.

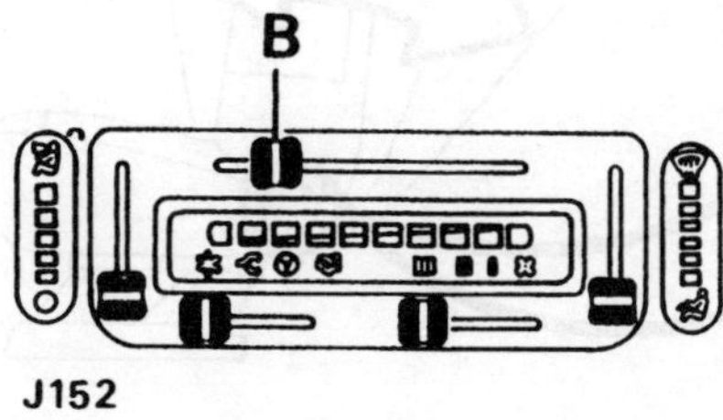

HIGHWAY DRIVING - Fig. J152

During a long journey when the temperature and humidity are extremely high and the air conditioning is being used, frost may form on the cooling coils of the evaporator. To prevent this, move the slider B slightly to the right, away from its extreme cold position.

Front seat adjustment

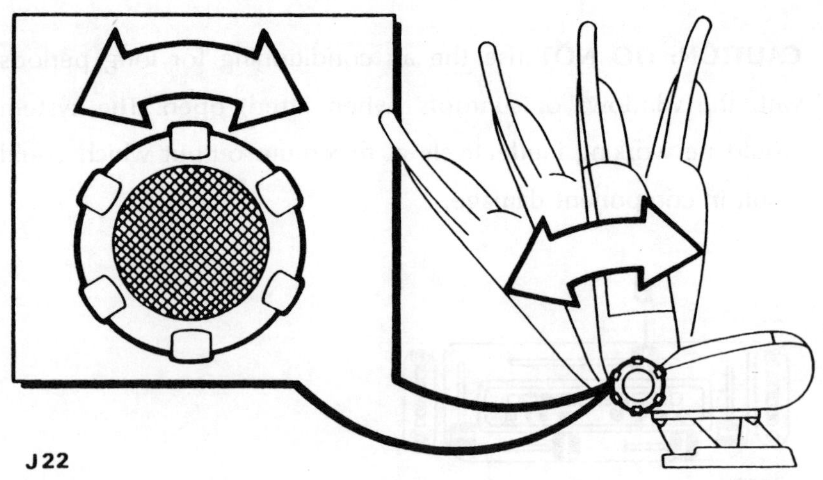

BACKREST MOVEMENT - Fig. J22

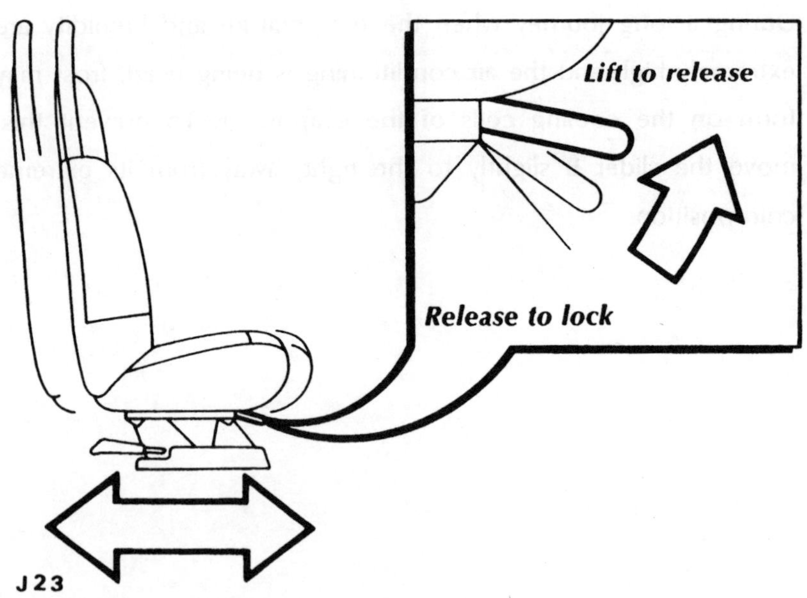

SEAT BASE MOVEMENT - Fig. J23

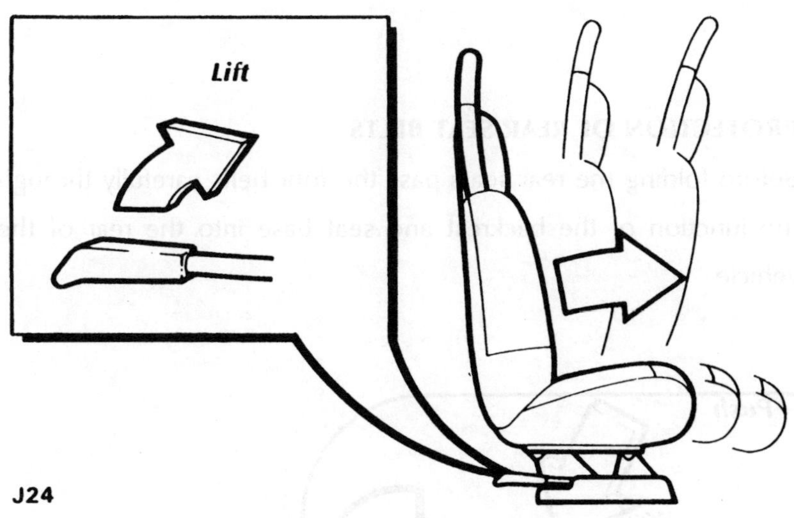

J24

TO TILT THE SEAT - Fig. J24

On some models the driver's seat is fixed.

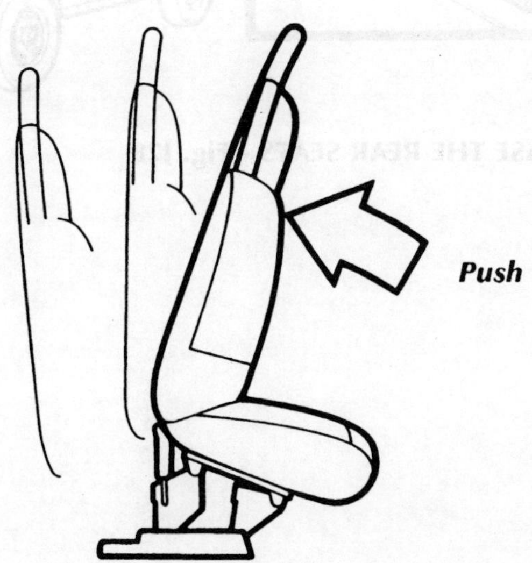

J 27

TO RETURN - Fig. J27

Rear folding seats

PROTECTION OF REAR SEAT BELTS

Before folding the rear seats pass the four belts carefully through the junction of the backrest and seat base into the rear of the vehicle

TO RELEASE THE REAR SEATS - Fig. J28

Folding rear seats

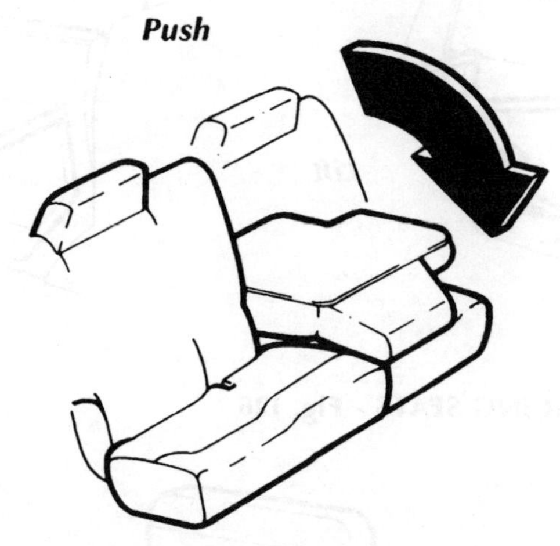

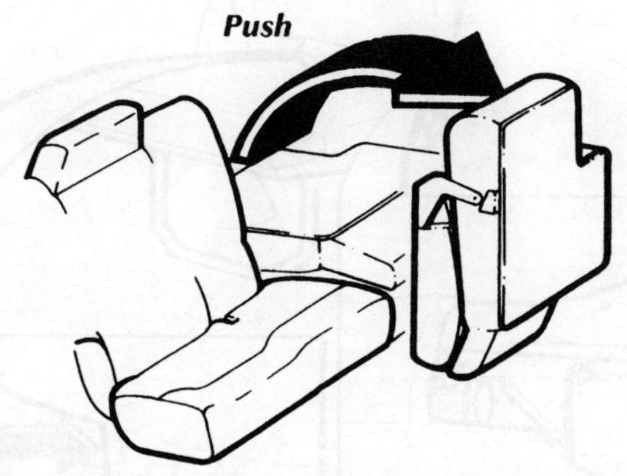

FOLD REAR SEATS - Fig. J25 and J30

Inward facing seats

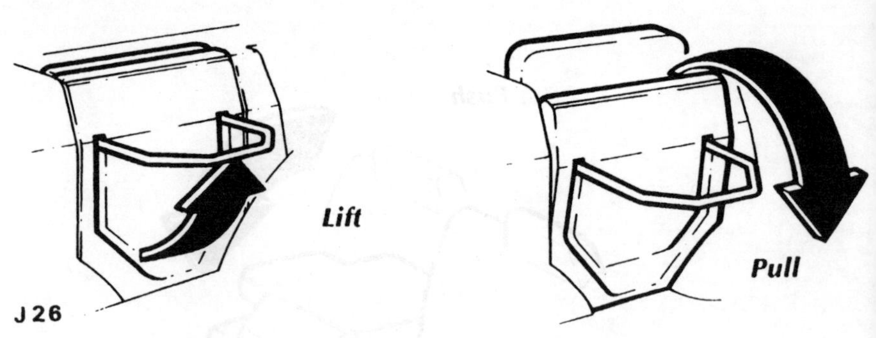

J 26

INWARD FACING SEATS - Fig. J26

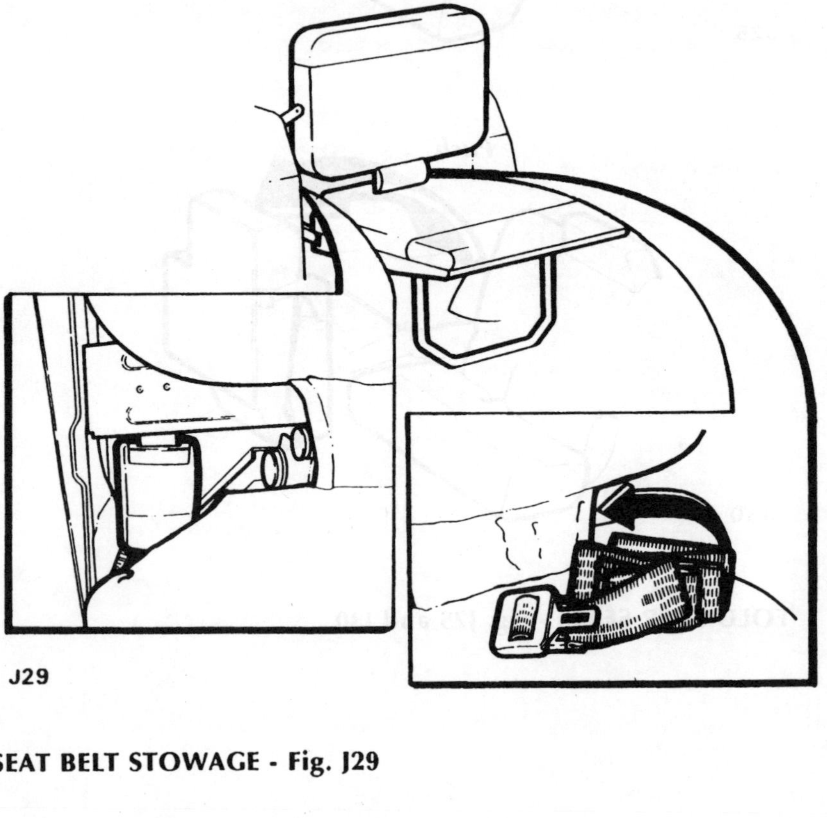

J29

SEAT BELT STOWAGE - Fig. J29

GENERAL SEAT BELT INFORMATION

The seat belts fitted to this vehicle are possible life-saving equipment, and should be regarded with the same importance as steering and brake systems.

⚠ **WARNING: Seat belts are designed to bear upon the boney structure of the body, and should be worn low across the front of the pelvis, or the pelvis, chest and shoulders, as applicable. Wearing the lap section of the belt across the abdominal area must be avoided.**

- Always use the seat belts provided, even for the shortest of journeys.
- Alterations and additions must not be made to the seat belts fitted to this vehicle.
- The seat belts must be fitted to the anchorage points provided.

When using the seat belt always ensure that the following points are observed.

- Always ensure that the belt is lying flat and is not twisted either on the wearer's body or between the wearer and the anchorage point.
- Never attempt to use the seat belt for more than one person.

Seat belts

Inertia

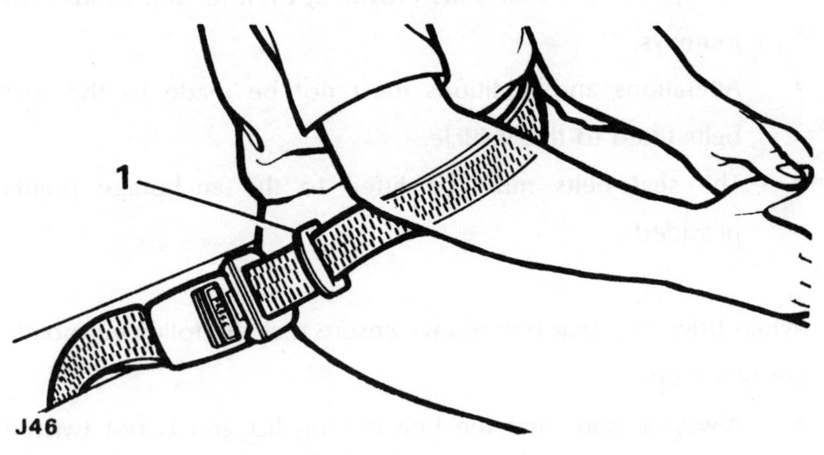

Lap strap

USING THE SEAT BELTS - Fig. J45 and J46

Two types of seat belt are in use, inertia reel (automatic) for the driver and outer passengers, and lap type for all other passengers.

To fasten, draw the tongue of the belt over the shoulder and across the chest, then push it into the engagement/release slot. A positive click indicates that the belt is locked.

To release, press the release button which will disengage the buckle. This allows the belt to retract.

To adjust lap type belts, slide the adjuster (1) along the belt and feed the webbing through the buckle until the belt is comfortably tight.

TESTING THE SEAT BELT

⚠ **WARNING: This test must be carried out under safe road conditions, i.e. level dry road with no following or oncoming traffic.**

With the belts in use, drive the vehicle at 8 kph (5 mph) and brake sharply. The automatic locking device should operate and lock the belt. It is essential that the driver and passenger are sitting in a normal relaxed position when making the test. The retarding effect of the braking must not be anticipated.

Seat belts

CARE OF THE SEAT BELTS

Inspect the belt webbing periodically for signs of abrasion and wear, paying particular attention to the fixing points.

If the vehicle has been involved in an accident, the seat belt assemblies must be replaced. If the webbing, buckle or cabling show any sign of wear or damage they must be replaced. If any fault is found report it immediately.

INFANT AND CHILD RESTRAINTS

When installing and using any infant or child restraint system, always follow the instructions provided by the manufacturer concerning its installation and use.

The failure to properly secure the child restraint system in the vehicle can endanger the child in the event of a collision or sudden stop and cause injury to other passengers. The centre rear seating position is fitted with lap belts which can be manually tightened to secure the infant or child restraint system. Older children should use the lap/shoulder belt.

SEAT BELT CLEANING

Do not attempt to bleach or re-dye the webbing. If the belts become soiled they should be sponged with warm water. Use a non-detergent soap not a caustic soap or chemical cleaner. Allow the belts to dry naturally not with artificial heat or by direct exposure to the sun.

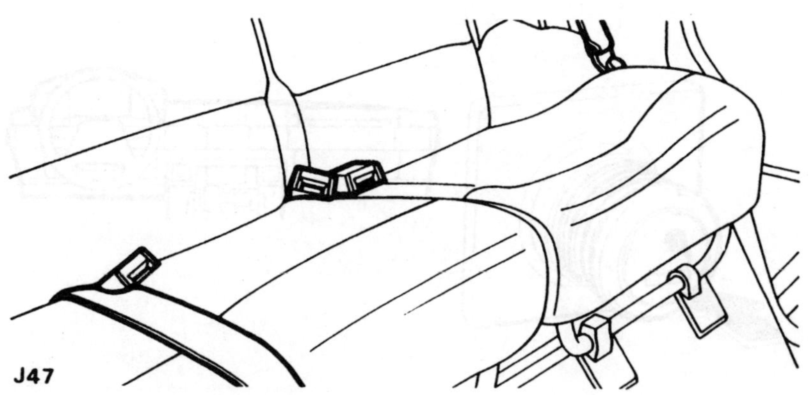

REAR SEAT BELT STOWAGE - Fig. J47

NOTE: Seat belts must be correctly stored as illustrated to prevent damage when using the rear of the vehicle.

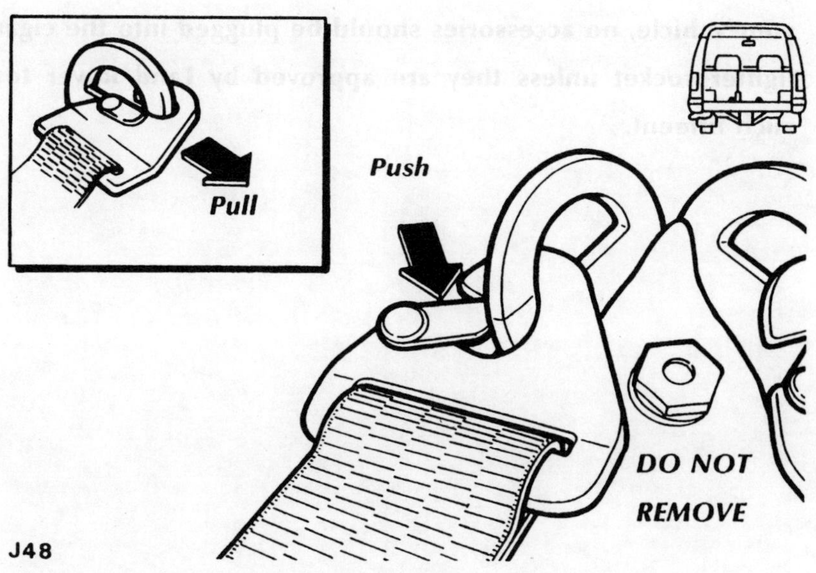

REMOVING REAR SEAT BELTS - Fig. J48

Cigar lighter

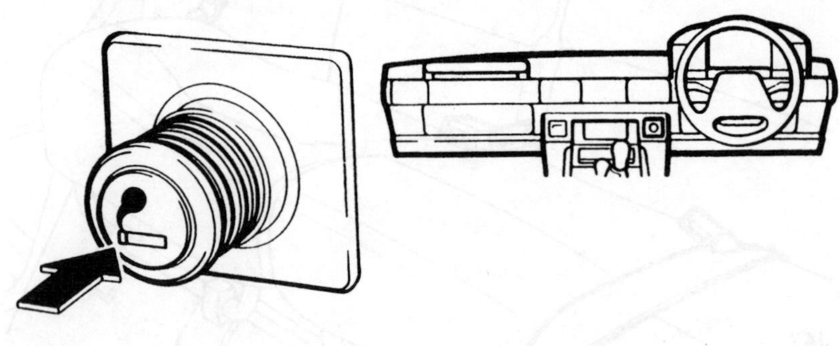

J156

CIGAR LIGHTER - Fig. J156

⚠ **WARNING: To avoid the risk of injury or damage resulting from overloading the electrical circuits on your vehicle, no accessories should be plugged into the cigar lighter socket unless they are approved by Land Rover for such fitment.**

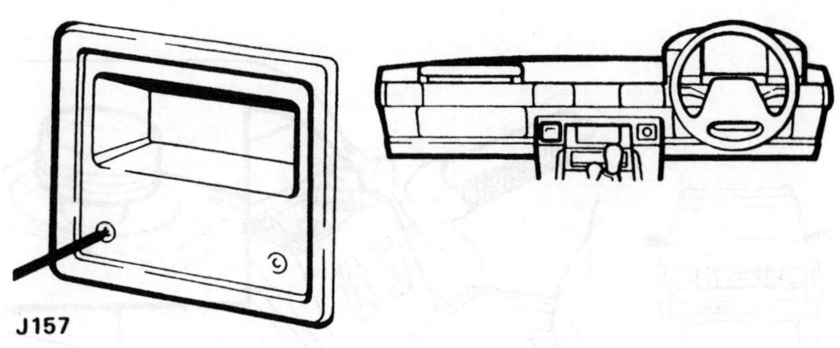

CLOCK - Fig. J157

To adjust the time, press the appropriate button until the correct figure is visible.

NOTE: The clock will need to be reset if the battery is disconnected.

Electric mirrors

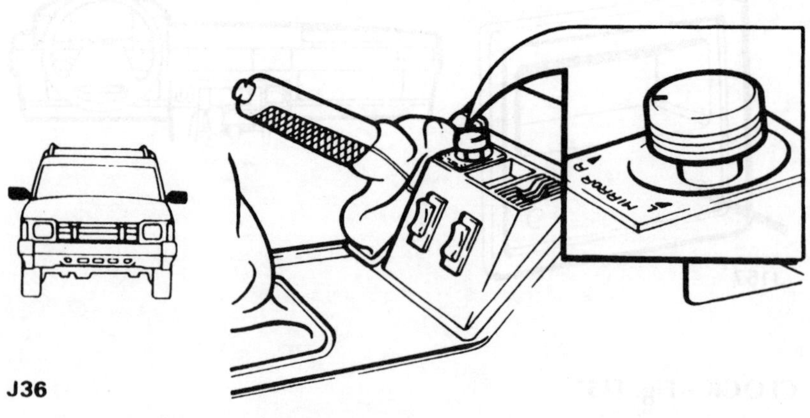

J36

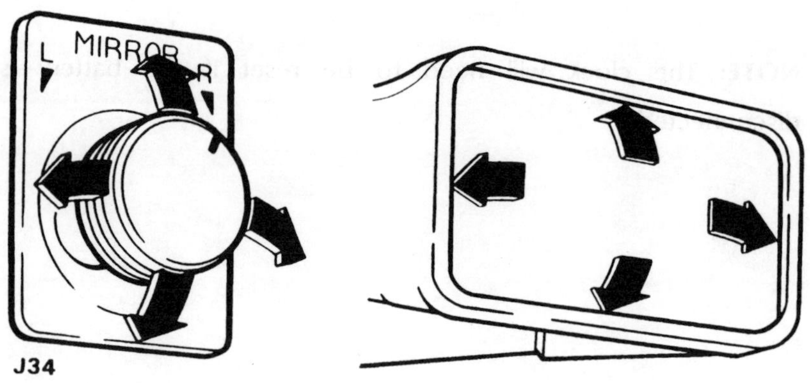

J34

ELECTRIC MIRRORS (when fitted) - Fig. J36 and J34

Turn the knob to "L" to adjust the left hand mirror, or to "R" to adjust the right hand mirror.

Door windows

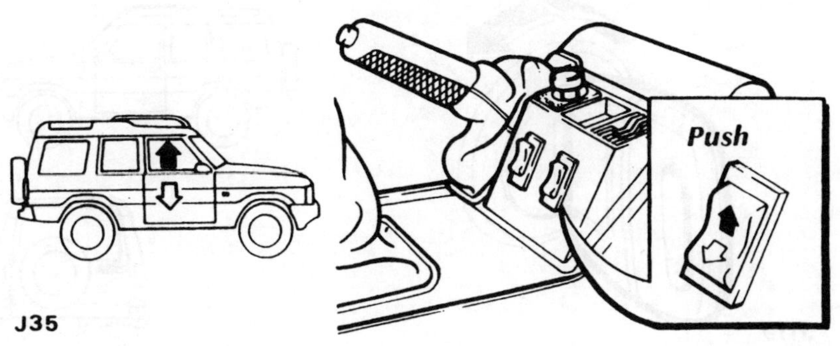

J35

ELECTRIC WINDOWS (when fitted) - Fig. J53

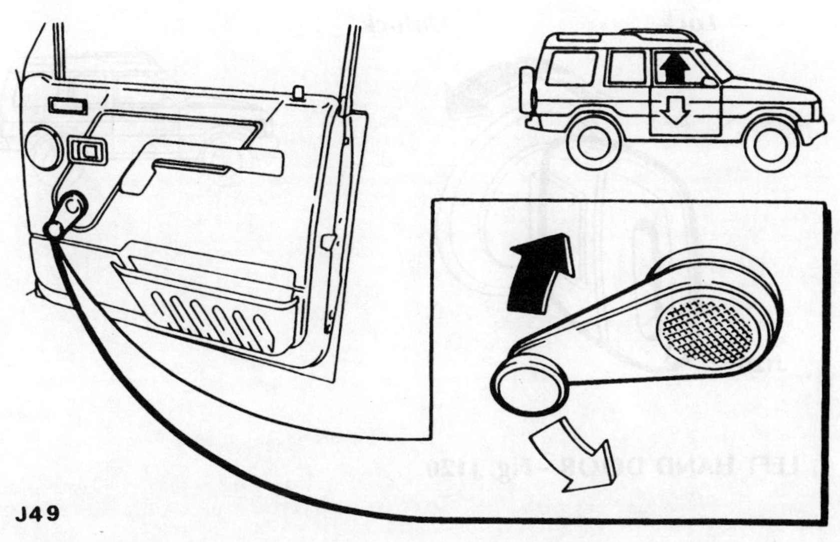

J49

WIND-DOWN WINDOWS - Fig. J49

Locking doors from outside

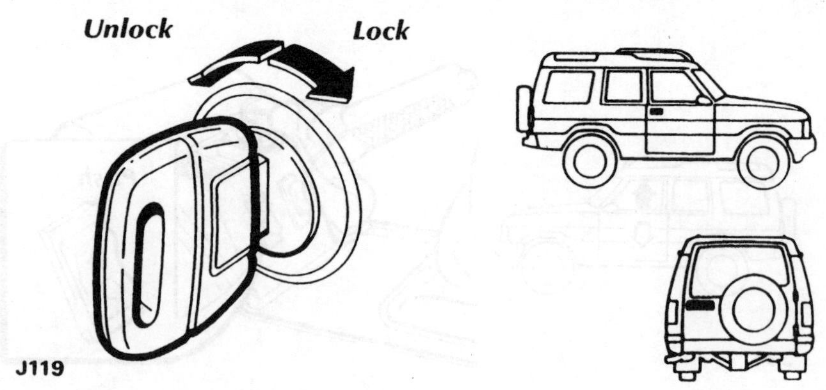

RIGHT HAND AND REAR DOOR - Fig. J119

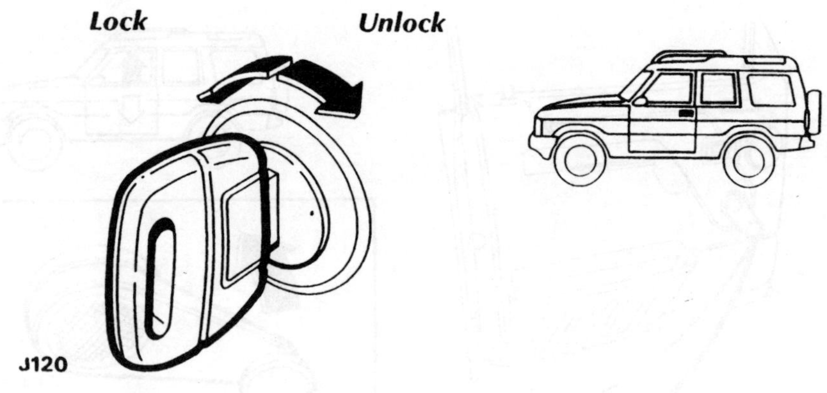

LEFT HAND DOOR - Fig. J120

CENTRAL DOOR LOCKING

By turning the key in the driver's door, all the doors will lock or unlock automatically.

Opening doors from outside

TO OPEN THE SIDE DOORS - Fig. J158

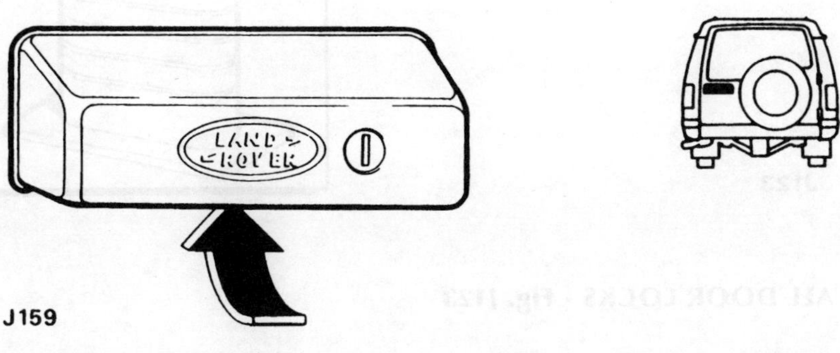

TO OPEN THE REAR DOOR - Fig. J159

Locking doors from inside

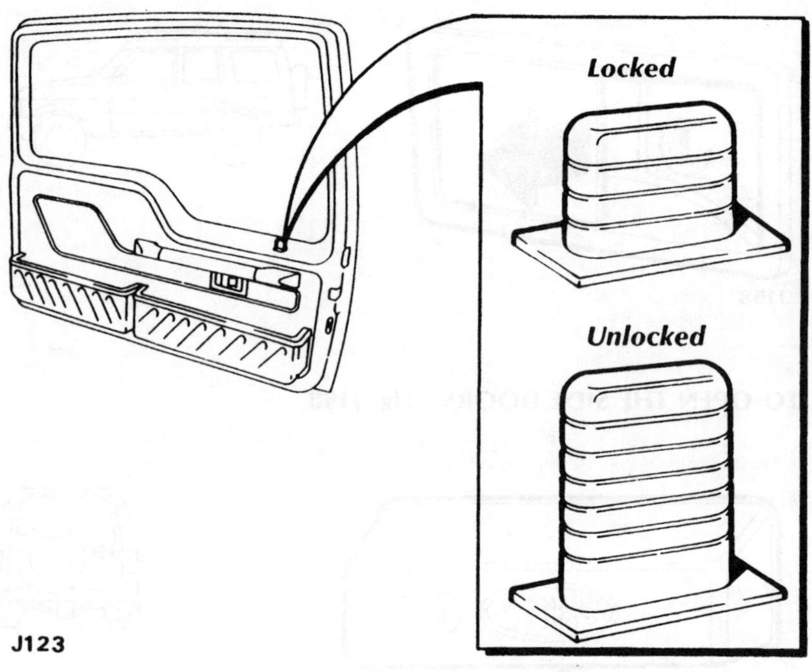

ALL DOOR LOCKS - Fig. J123

NOTE: The button can not be used to lock the side doors when leaving the vehicle. Use the key from the outside.

CENTRAL LOCKING

Using the button on the driver's door will lock or unlock all the doors. The rear and passenger doors can also be locked or unlocked independantly.

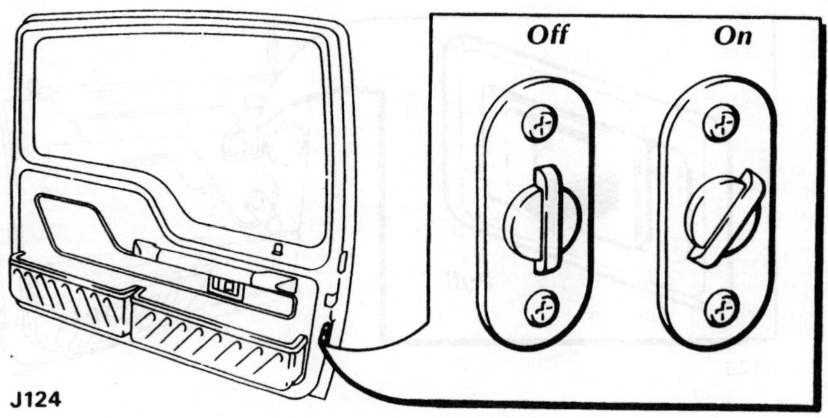

CHILD LOCK OPERATION - Fig. J124

With the child lock "on", the rear door can not be opened from the inside. With the child lock "off", the door can be opened from the inside as normal.

Opening doors from inside

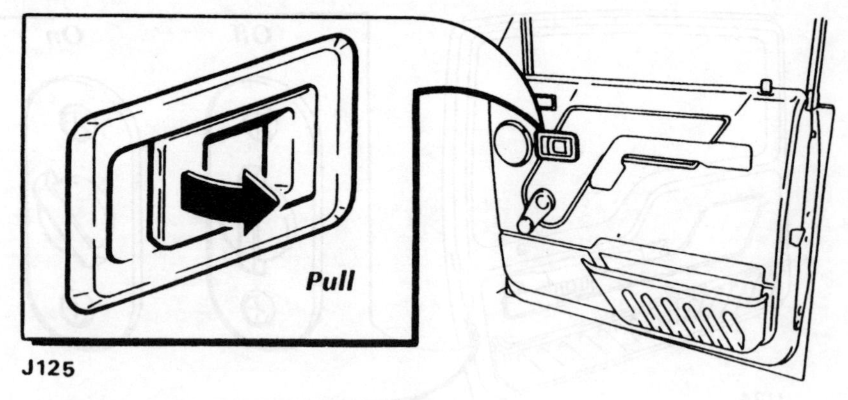

DOOR HANDLE - Fig. J125

All of the door handles will work in the same way.

Fuel filler

J126

UNLOCKING THE FUEL FILLER FLAP - Fig. J126

J127

LOCKING THE FUEL FILLER FLAP - Fig. J127

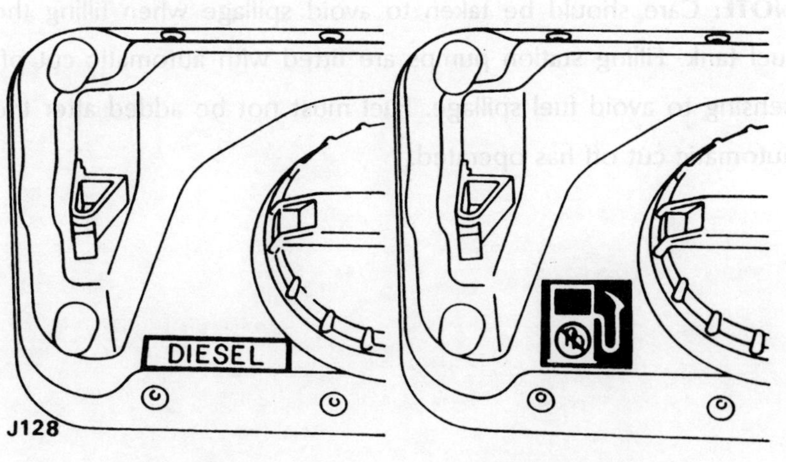

J128

Diesel fuel   *Unleaded fuel*

Fuel filler

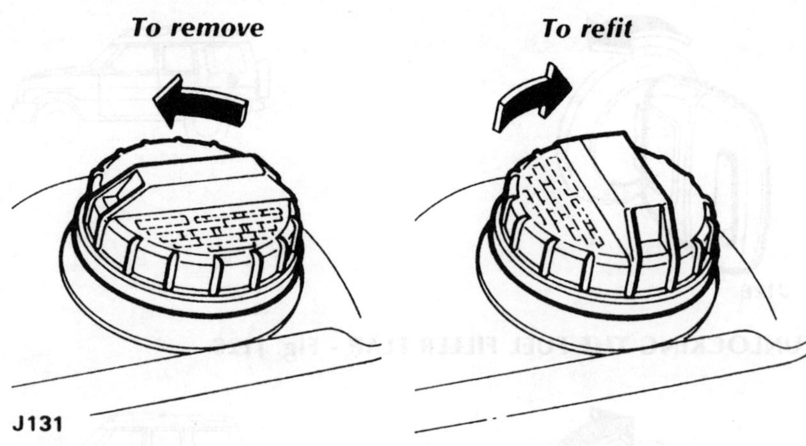

FUEL FILLER CAP - Fig. J131

CAUTION: To avoid sudden fuel discharge, rotate cap one half turn to vent. After hissing stops, continue turning to remove.

NOTE: Care should be taken to avoid spillage when filling the fuel tank. Filling station pumps are fitted with automatic cut off sensing to avoid fuel spillage. Fuel must not be added after the automatic cut off has operated.

Bonnet catch

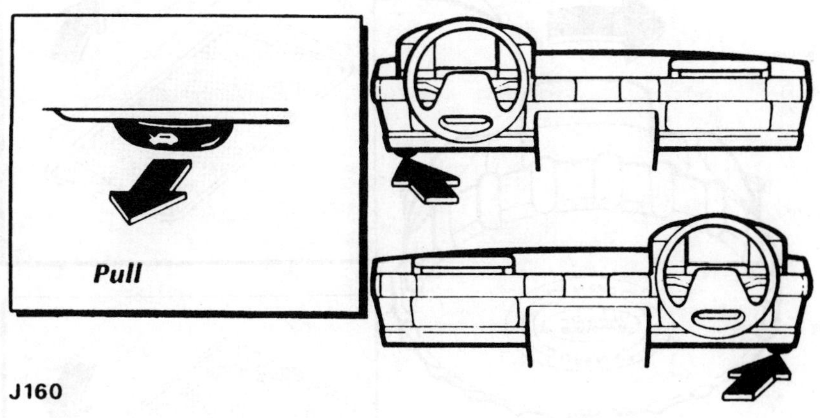

TO RELEASE THE BONNET - Fig. J160

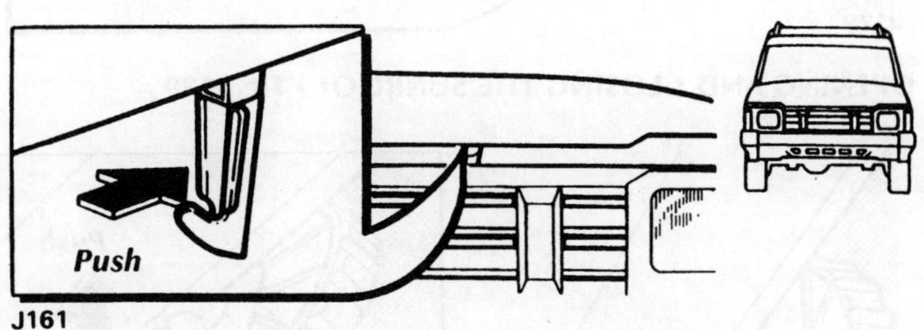

TO RELEASE THE SAFETY CATCH - Fig. J161

Sunroof (when fitted)

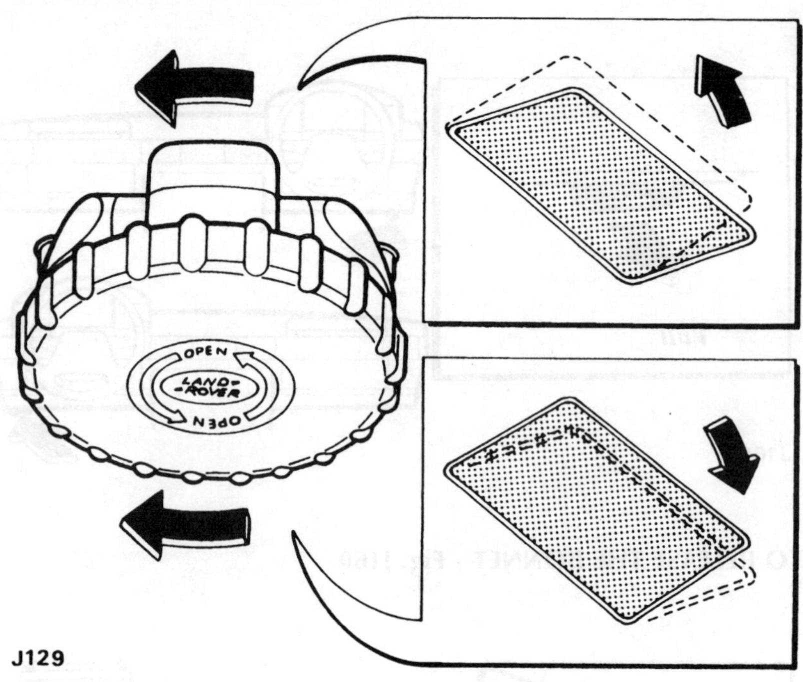

J129

OPENING AND CLOSING THE SUNROOF - Fig. J129

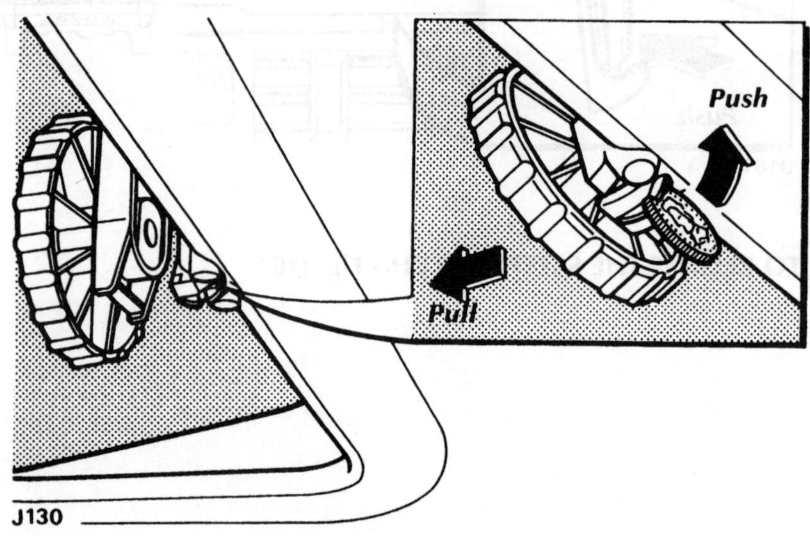

J130

REMOVING FRONT SUNROOF GLASS - Fig. J130

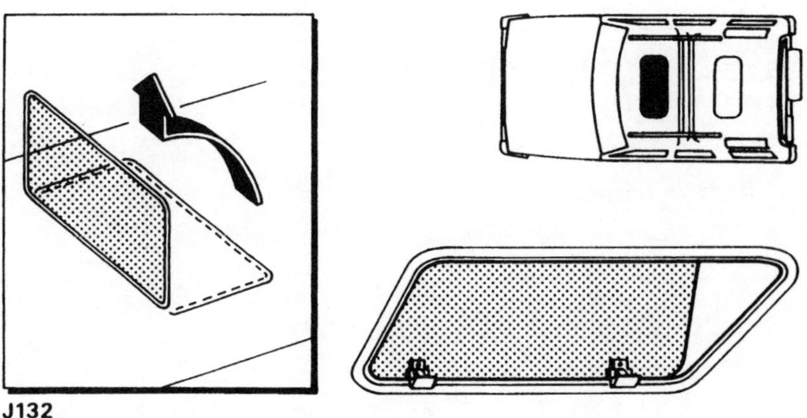

J132

REMOVING FRONT SUNROOF GLASS - Fig. J132

NOTE: The sunroof must be in the fully open position to expose the slot shown in J130. Only the front sunroof glass is removable. It must be stowed in the bag which is attached to the back of the rear seat.

DRIVING AND OPERATING

3

Low emission exhaust system

DRIVING AND OPERATING

LOW EMISSION EXHAUST SYSTEM

The Discovery V8 petrol engine already conforms to E.C.E. exhaust emission regulations, but for anyone wishing to further enhance performance, a catalyst low emission exhaust system is available. If your vehicle is fitted with a low emission exhaust system, or if you intend purchasing a conversion kit from Land Rover Parts, the following important points should be noted.

Always use **UNLEADED** (95 octane) petrol.

Note the cautionary and instruction labels fitted to the vehicle, specifying the use of **UNLEADED** fuel only.

NOTE: NEVER put any **LEADED** petrol in the fuel tank, as it would completely destroy the emission reducing properties of the catalyst exhaust.

Starting the engine - 200Tdi

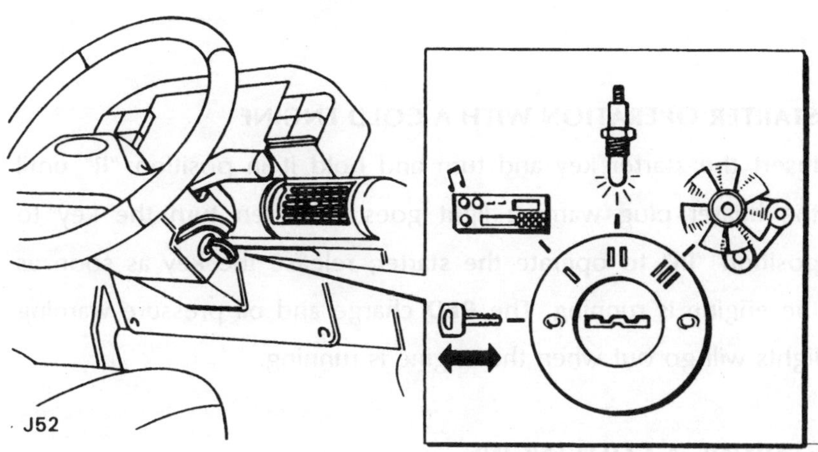

STARTING AND STOPPING - 200Tdi ENGINE - Fig. J52

The following procedures must be used to ensure easy starting and avoid damage to the turbo-charger.

Before starting the engine for the first time each day, check that the engine oil and radiator coolant levels are correct, top up if necessary. Check that the handbrake is on and that the main gear lever is in neutral.

HEATER PLUGS

The Discovery 200Tdi engine will start satisfactorily, with the proper use of the heater plugs, down to temperatures of -32°C (-22°F) even with batteries only 80% charged, provided the correct grade of oil is used. Turn the starter key to the heater plug position (II) when starting from cold.

An amber warning light will glow when the engine starter key is turned to the 'heater plug' position, and will go off after a few seconds when the starting temperature is correct.

Starting the engine - 200Tdi

STARTER OPERATION WITH A COLD ENGINE

Insert the starter key and turn and hold it in position "II" until the heater plug warning light goes off, then turn the key to position "III" to operate the starter; release the key as soon as the engine is running. The **RED** charge and oil pressure warning lights will go out when the engine is running.

STARTING A COLD ENGINE

DO NOT use the accelerator pedal during the engine starting procedure; extra fuel for cold starting is automatically supplied by the injector pump.

CAUTION: The engine must not be run above fast idle until the oil pressure warning light goes off; this is to ensure that the engine bearings and the turbo-charger bearings are receiving lubrication before being run at speed.

In cold weather, depress the clutch pedal while the starter motor is in operation to improve engine starting speed.

DO NOT operate the starter for longer than 10 seconds; if the engine fails to start, switch off and wait 10 seconds before re-using the starter. If after a few attempts the engine fails to start, switch off and investigate the cause.

CAUTION: Continued use of the starter will not only discharge the battery but may damage the starter motor.

STARTING A WARM ENGINE

DO NOT operate the accelerator pedal during the engine starting procedure. Turn the starter key to the engine start position. Release the key immediately the engine starts.

STOPPING THE ENGINE

To avoid the possibility of inadequate lubrication of the turbo-charger, the following precaution must always be observed.

- Before stopping the engine, allow it to idle for 10 seconds to give time for the turbo-charger to slow down whilst oil pressure is available at the bearings.

- Switching the engine off too quickly could leave the turbine rotating at several thousand revolutions per minute without oil pressure.

Starting the engine - 200Tdi

PRECAUTIONS FOR COLD WEATHER PROTECTION

The following recommendations should be considered to minimize difficulties associated with cold weather fuel problems.

- Ensure 'winter' grade diesel fuel is used. Filling stations should automatically change to this fuel during winter.

- Renew the main fuel filter element at the recommended intervals.

- Maintain the state of charge of the battery in a satisfactory condition.

- Follow the starting procedures stated.

The use of paraffin (kerosene) as a diesel fuel additive, is illegal in the U.K. and the use of petrol as a fuel in a diesel engine is highly dangerous.

WARNING: Diesel exhaust fumes contain noxious substances which can cause unconsciousness and may even be fatal. Do not breathe exhaust gas because it contains noxious substances which by themselves have no colour or odour. Never start or leave the engine running in an enclosed unventilated area. If you think exhaust fumes are entering the vehicle have the cause determined and corrected immediately.

Starting the engine - petrol

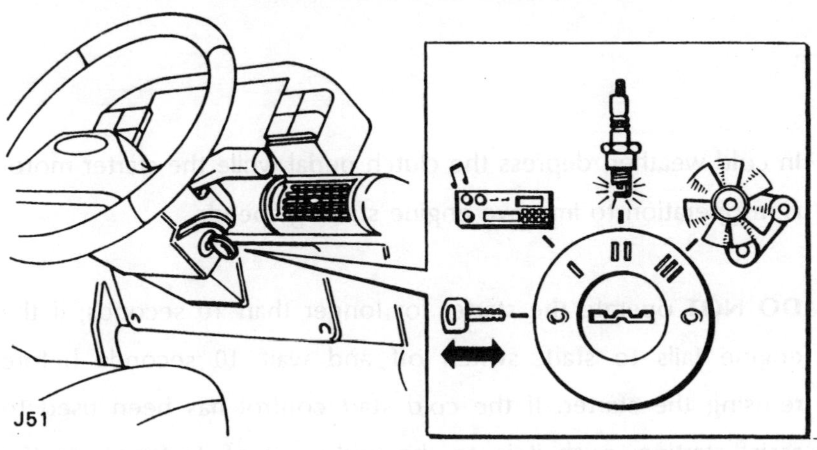

STARTING - PETROL ENGINE - Fig. J51

Before starting the engine for the first time each day, check that the engine oil and radiator coolant levels are correct, top up if necessary. Check that the handbrake is on and that the main gear lever is in neutral. If the engine is cold, use the cold start control as follows:

Depress the accelerator pedal half way down. Pull out the cold start control while the accelerator is depressed, then release the accelerator.

STARTER OPERATION

Insert and turn the ignition key to position "II", then turn the key to position "III" to operate the starter; release the key as soon as the engine is running. The **RED** ignition and oil pressure warning lights will go out when the engine is running.

Starting the engine - petrol

In cold weather, depress the clutch pedal while the starter motor is in operation to improve engine starting speed.

DO NOT operate the starter for longer than 10 seconds; if the engine fails to start, switch off and wait 10 seconds before re-using the starter. If the cold start control has been used to assist starting, push it in to the mid-point of its travel. As the engine warms and runs smoothly, progressively return the control to the fully in position. If after a few attempts the engine fails to start, switch off and investigate the cause.

CAUTION: Continued use of the starter will not only discharge the battery but may damage the starter motor.

STARTING WITH A WARM ENGINE
DO NOT use the cold start control or pump the accelerator pedal. Depress the accelerator pedal to approximately half-way. Turn the ignition key to operate the starter, keeping the accelerator pedal in the half-way position. Release the ignition key and accelerator pedal immediately the engine starts.

Starting the engine - petrol

⚠️ **WARNING: Carbon monoxide is a dangerous gas and can cause unconsciousness and may even be fatal. Do not breathe exhaust gas because it contains carbon monoxide which by itself has no colour or odour. Never start or leave the engine running in an enclosed unventilated area. If you think exhaust fumes are entering the vehicle have the cause determined and corrected immediately.**

Engine fault finding

ENGINE STARTING FAULT FINDING

- PETROL AND DIESEL ENGINES

This chart gives a number of quick and easy checks which may help you to get the engine started. Checking of any part of the electronic ignition system must be referred to your Land Rover Dealer or Distributor.

FAULT	POSSIBLE CAUSE
STARTER MOTOR WILL NOT OPERATE If the starter motor does not operate when the key is turned to the start position the fault may be -	1. Loose battery lead 2. Discharged battery 3. Faulty wiring 4. Faulty starter solenoid 5. Faulty starter motor
STARTER MOTOR OPERATES BUT ENGINE WILL NOT START Do not use the starter motor for long periods, if the engine will not start, the fault may be -	1. Lack of fuel 2. Fuel not reaching engine 3. Petrol engines-ignition fault Diesel engines heater plug fault

> **CAUTION:** If the vehicle runs out of fuel or engine will not start, turn off the ignition/starter switch to prevent damage to electrical components.

⚠ **WARNING: The electronic ignition system on petrol models involves very high voltages. Inexperienced personnel and wearers of medical pacemaker devices should not be allowed near any part of the high-tension circuit.**

CHECK

1. Clean and tighten the battery lead connections and try starting the engine.
2. Try push or tow starting the vehicle.
3. Seek expert advice.
4. Seek expert advice.
5. Seek expert advice.

---

1. The fuel gauge could be faulty. Check that there is fuel in the tank.
2. a Slacken a fuel pipe nut at the engine and operate the starter for a few seconds. If fuel has leaked from the slackened nut, then fuel is reaching the engine. Retighten the fuel pipe nut and wipe away any fuel.
   b If fuel has not leaked, check for broken or disconnected pipes.
   c If the fault is not found, seek expert help.
3. **PETROL ENGINES:** Remove one spark plug. Reconnect its lead and place it on the engine so that there is a metal-to-metal contact. Operate the starter while someone else looks to see if the plug sparks. If a spark is seen, ignition fault is unlikely. Refit the plug and make sure that the leads are connected in the right order. If no spark is seen, refit the plug, as above, and make sure that all the leads are firmly connected to the distributor cap, and that the centre lead is firmly connected to the ignition coil. Try starting the engine again. If the engine does not start, remove the distributor cap and wipe the inside to remove any damp or dirt. Refit the distributor cap and try starting the engine again. If the engine does not start, seek expert help.
   200Tdi ENGINES: Make sure that the yellow/black leads are firmly connected to all four heater plugs, and that the black/yellow lead is firmly connected to the rear heater plug. Hold the starter key in the 'heater plug' position for at least twelve seconds, then try to start the engine. If the engine does not start, seek expert help.

Day-to-day running

DRIVING - PETROL AND DIESEL MODELS

WARMING-UP

When the engine is cold, drive the car as soon as the engine has started.

DO NOT warm-up the engine by running it at a slow speed with the vehicle stationary.

CAUTION: Harsh acceleration and labouring the engine before normal temperature is reached can damage the engine.

EXCESSIVE TEMPERATURE

Excessive engine temperature is indicated when the temperature gauge indicator reaches the **RED** graduations. Any sudden increase in engine temperature must be investigated. Stop the engine and check the engine cooling system.

WARNING: Do not remove the filler cap from the expansion tank or the filler plug from the radiator when the engine is hot because the cooling system is pressurised and personal scalding could result.

Ensure there are no leaks, top up the radiator expansion tank if necessary. Make sure the fan belt is not broken and is correctly tensioned.

DRIVING CHARACTERISTICS

⚠ **WARNING: Your Discovery has a higher ground clearance and hence a higher centre of gravity than an ordinary passenger car to enable it to perform in a wide variety of off-road applications. An abrupt manoeuvre at an inappropriate speed or on an unsuitable surface could cause the vehicle to go out of control.**

RUNNING-IN PERIOD

Progressive running-in of a new Discovery is important and has a direct bearing on reliability and smooth running throughout its life.

The most important point is **NOT** to hold the vehicle on large throttle opening for any sustained periods. To start with, the maximum speed should be limited to 65 to 80 km/h (40 to 50 mph) for diesel models and 80 to 95 km/h (50 to 60 mph) for petrol models, on a light throttle and this may be progressively increased over the first 2.500 km (1,500 miles).

FUEL RECOMMENDATIONS

Recommended fuels for petrol models are specified in the Data section. No advantage will be gained by the use of higher octane fuels, than those recommended.

Day-to-day running

UNLEADED PETROL

All current V8 petrol engines used in Discovery can be run on unleaded or leaded petrol. It is strongly recommended, that whenever it is available, unleaded petrol should be used to help protect the environment. It is permissible to mix unleaded and leaded petrol when refilling the petrol tank.

CAUTION: Do not use oxygenated fuels such as blends of methanol/gasoline or ethanol/gasoline (e.g. Gasohol).

WARNING: Do not fill the fuel tank completely if the vehicle is to be parked in direct sunlight or high ambient temperature, as this would cause the fuel to expand and escape through the breather pipe onto the ground.

200Tdi ENGINES

Clean, good quality fuel should be used in diesel models. It is important that the sulphur content of diesel fuel does not exceed 1%. In Europe all supplies should be within the limit, but in other areas operators should check with their suppliers. Change the fuel filter element at the recommended service intervals and clean the sediment bowl regularly.

BRAKE, SERVO ASSISTANCE AND POWER ASSISTED STEERING

DO NOT coast in neutral with the engine switched off as the brake servo and power assistance for the steering will not operate. The brakes and steering will still function but more effort will be required by the driver. This will also apply if the vehicle is being towed without the engine running, and extra caution must be used.

SNOW CHAINS

Chains may be fitted to provide increased traction during heavy snow conditions. Never fit chains to one wheel only, always fit snow chains in pairs and to the rear axle only. Ensure the gearbox differential control is in the **LOCKED** position. Remove the snow chains immediately the road is clear of snow.

VEHICLE HEIGHT

Always check the headroom before driving through low entrances to ensure there is sufficient clearance. This is particularly important if the vehicle is fitted with a roof rack or sunroof, which is open.

Driving in general

DRIVING IN GENERAL

⚠ **WARNING: Always wear a seat belt for personal protection while either ON-ROAD or OFF-ROAD driving. Driving off-road can be particularly hazardous therefore do not take risks. Drive carefully.**

⚠ **WARNING: DO NOT apply the handbrake while the vehicle is in motion as this could result in loss of vehicle control and damage to the transmission.**

⚠ **WARNING: DO NOT rest your foot on the brake pedal while travelling as this may overheat the brakes, reduce their efficiency and cause excessive wear.**

⚠ **WARNING: DO NOT wrap your thumbs round the steering wheel as any steering kick-back over rough ground may result in personal injury.**

⚠ **WARNING: DO NOT coast with the engine switched off as the brake servo and steering assistance will not operate. The brakes will still function but more foot pressure will be required.**

Driving in general

⚠ **WARNING: Driving through water or heavy rain will result in braking surfaces becoming coated with moisture. This will affect braking efficiency until the surfaces are dried by intermittent light application of the brakes which should be done at a safe distance from other vehicles. Brakes should be dried and tested immediately after driving through water, every few miles when driving in heavy rain and especially before leaving a wet motorway. When parking, do not rely on the handbrake alone to hold the vehicle if the brake linings have been subjected to immersion in mud and water (See 'After wading' details later in this section).**

DO NOT use a gear which is too high for the vehicle speed and travel conditions involved; it is preferable to select a lower gear and use more revolutions rather than allow the engine to labour at low speed.

DO NOT use the clutch pedal as a foot rest. Keep the left foot well clear of the clutch pedal while the vehicle is in motion.

DO NOT overload the vehicle for sustained cross country work. See 'Vehicle weights' Section 5.

Towing

J73

TOWING

Consult a Land Rover Dealer for details and advice on approved towing equipment and accessories.

Discovery models can tow loads over various types of terrain.

The torque ranges of Discovery engines allow maximum-weight loads to be driven smoothly from standstill, and reduces gear changing on hills, or rough terrain. A smooth start will be achieved with trailers over 2000 kg (4400 lb) by moving off in low range then changing to high range while on the move.

The suspension is designed to cope with a heavy trailer load without upsetting the balance or feel of the vehicle. Details of gross maximum trailer weights are listed on the following page.

When preparing the vehicle and trailer combination, careful attention must be paid to the trailer manufacturer's recommendations. An outline of the correct procedure is given here:

(a)   Vehicle tyre pressures must always be set at normal pressures, See Data Section.

(b)   Adjust tyre pressures on the trailer, as recommended by the manufacturer.

(c)   Balance the trailer and the vehicle, both unladen, so that the trailer draw-bar and the hitch point on the vehicle are at the same height. Adjust the height of the hitch point if necessary.

(d)   Check operation of trailer brakes and lights.

(e)   Load the trailer and check the weight on the hitch point (called the drawbar loading weight, or nose weight), in accordance with the manufacturer's recommendations.

(f)   The recommended nose weight limit is 75 kg (165 lb). The nose weight plus the load area and/or rear seats of the vehicle together with load and passengers must never exceed maximum rear axle load.

The following points must also be allowed for when loading a trailer.

(a)   The weight of the trailer plus load.

(b)   Towing stability.

(c)   Weight of the vehicle contents including passengers. When part of the weight is transferable, loading the towing vehicle will generally improve the stability of the combination.

(d)   Altitude: Engine performance is progressively reduced above altitudes of 300 m (1,000 feet).

Towing

NOTE: Since towing regulations vary from country to country, it is important to refer to the relevant national motoring organisations for the laws relating to towing weights and speed limits. The following maximum permissible towed weights refer to the design limitations.

Maximum Permissible Towed Weights	On-road kg	Off-road kg
Unbraked trailers	750	500
Trailers with overrun brakes	3500	1000
4 wheel trailers with coupled brakes *	4000	1000

NOTE: * In order to tow a trailer with a weight in excess of 3,500 Kg, it is necessary to adapt the vehicle to operate a Coupled Brake System, and the VIN plate must be changed to show the increased train weight.

A revised VIN plate may be obtained from Land Rover, which will be issued, subject to satisfactory proof that the vehicle has been fitted with an approved conversion.

Roof rack (if fitted)

ROOF RACK - Fig. J74 and J117

An approved front roof rack is available as a factory fitted option on estate models.

The roof rack has two side rails which are permanently fixed to the roof of the vehicle and three removeable cross rails which are stowed in the tool bag beneath the rear seat.

Each cross rail has a pair of locating pins at each end.

Roof rack (if fitted)

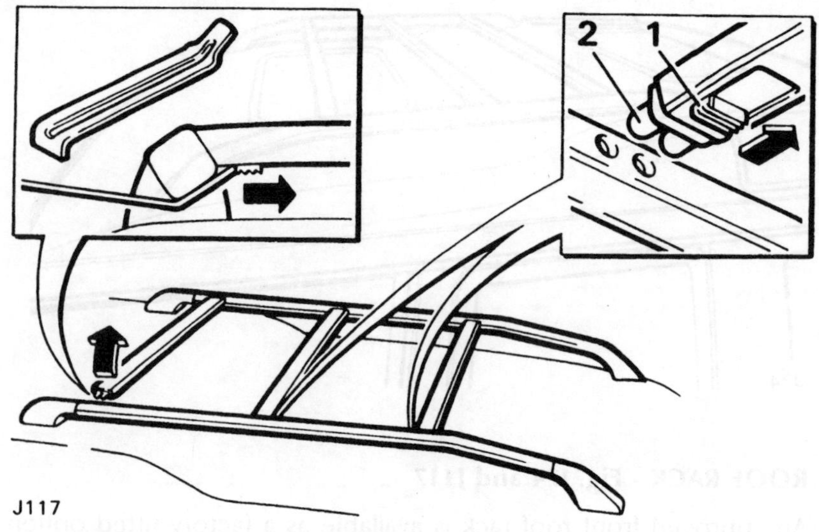

Fit the cross rails with the grooved side facing up, by engaging the locating pins in the holes provided in one side rail, slide the button (1) inwards, locate the spring loaded locating pins (2) in the holes in the opposite side rail and release the button. Check that the pins are correctly located and that the cross rails are secure.

NOTE: If any difficulty is experienced in fitting or removing the cross rails, use the special tool supplied in the vehicle tool kit to operate the spring loaded catch.

Roof rack (if fitted)

MAXIMUM roof rack load is 50kg evenly distributed.

The load must not overhang or exceed the perimiter of the rack. For longer loads such as surfboards or ladders, a long rack which attaches to the rear roof is available from Land Rover Parts; and can be purchased from your Land Rover dealer.

DO NOT fit unapproved roof racks as they will not have been tested by Land Rover for security and load carrying, and could lead to damage or other problems.

WARNING: When the use of a roof rack is required care should be taken as the vehicle stability will be affected. The roof rack load MUST be evenly distributed and firmly secured to the side rails NOT JUST to the cross members. Drive carefully and be prepared for any emergency.

Roof rack (if fitted)

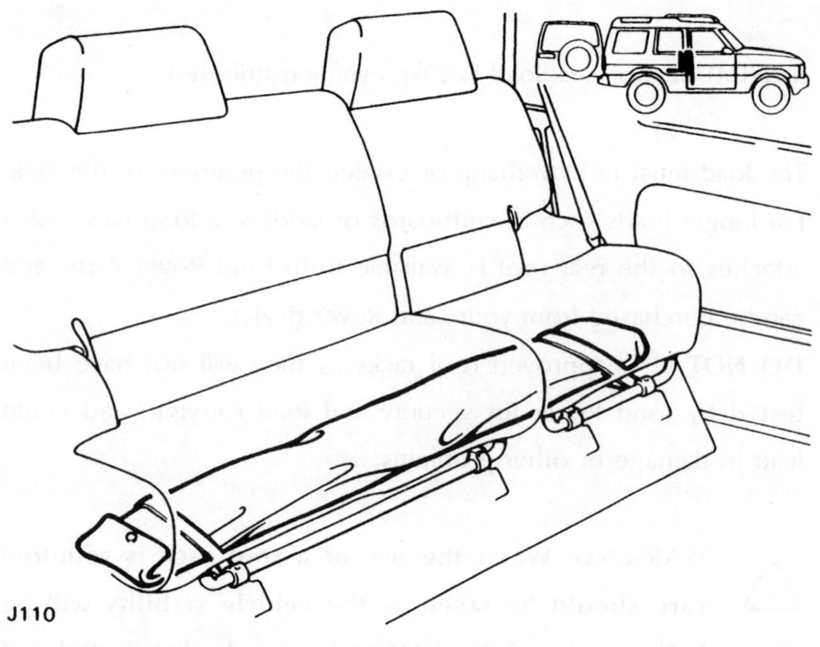

J110

When the roof rack is not in use, the cross rails can be removed and securely stowed in the tool bag beneath the rear seat.

⚠️ **WARNING: DO NOT leave the roof rack cross members loose in the vehicle, as they could move and cause personal injury in the event of an accident or emergency manoeuvre either on or off road.**

Accessories and conversions

ACCESSORIES AND CONVERSIONS

Discovery vehicles are designed and constructed for a variety of uses but no alterations or conversions should be carried out to any vehicle produced by Land Rover which could affect the safety of the vehicle or its passengers.

Land Rover has tested and approved a large number of accessories and conversions, suitable for Discovery in its current range. Before fitting **ANY** accessory or commencing **ANY** conversion work to any Discovery vehicle, **CHECK** that the accessory or conversion is approved by Land Rover.

WARNING: DO NOT FIT unapproved accessories or conversions, as they could affect the safety of the vehicle. Land Rover will not accept any liability for death, personal injury or damage to property which may occur as a direct result of fitment of non-approved accessories or the carrying out of non-approved conversions to Discovery vehicles.

CAUTION: DO NOT use auxiliary devices, such as roller generators, that are driven by one wheel of the vehicle, as they could cause failure of the gearbox differential. If the gearbox differential lock is engaged in an attempt to avoid damage, the vehicle will drive itself forward.

How to use the gearboxes

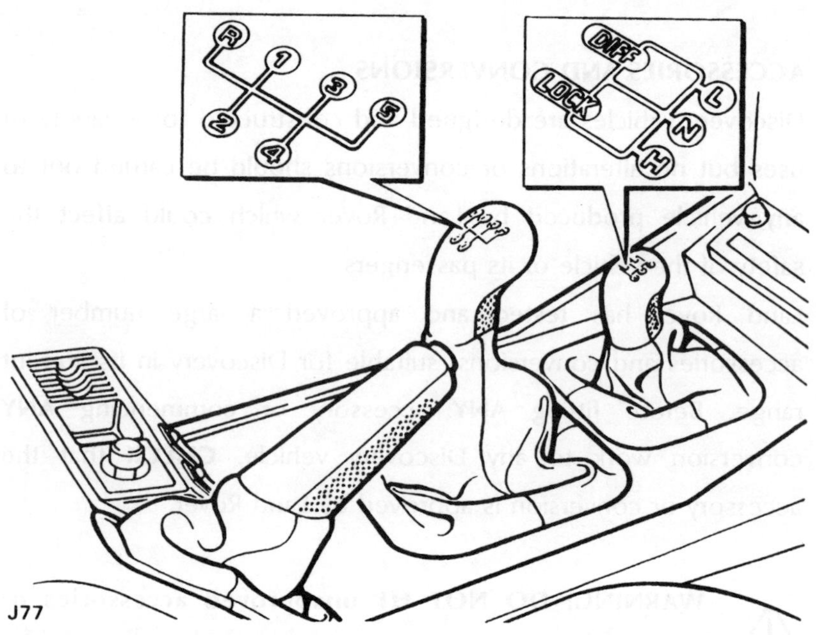

DRIVING TECHNIQUES

All Discovery vehicles have a five speed manual gearbox coupled
to a two speed transfer box, producing ten forward gears and
two reverse gears. A central differential in the transfer box
distributes the drive to the front and rear axles, providing
permanant four wheel drive.

The main gearbox is operated in the normal way with the gear
positions marked on the gear lever knob. In neutral, the gear
lever is automatically aligned with the third/fourth gear positions
and reverse is protected against accidental selection by spring
pressure. Note that fifth gear is designed to reduce engine
speed and improve economy when cruising.

How to use the transfer gearbox

The second gearbox, known as the transfer box allows the driver to select a high or low range of gears. The box is operated by the transfer gear lever which also controls the central differential **'DIFF LOCK'**.

High range (H)

Use high range for all normal road driving and for dry level off-road driving.

Low range (L)

In any situation where low speed manoeuvering is necessary, such as reversing a trailer or negotiating a boulder strewn river bed, use low range in extreme off-road conditions where progress in high range could not be maintained.

'DIFF LOCK' Central differential

Use the 'unlocked' position for all normal driving to allow the central differential in the transmisson to smooth the drive to the front and rear axles. Use the **'DIFF-LOCK'** position to improve traction in extreme conditions where wheel grip may be lost, such as wet grass, mud, sand, ice or snow. Return to the 'unlocked' position as soon as good, dry and firm ground is reached. **DO NOT** use the **'DIFF LOCK'** unnecessarily, as this would cause stress in the transmission and eventual damage.

How to use the transfer gearbox

USING THE TRANSFER BOX

All normal driving should be done in high range (H) without any need to use the transfer gear lever. If driving conditions are encountered where low range (L) or **'DIFF LOCK'** is required, use the transfer gear lever, as follows.

NOTE: There are two ways of operating the transfer lever, the 'normal' method is to help new drivers to make smooth, quiet gear changes before using the 'advanced' method for experienced drivers.

NORMAL METHOD

Changing from high to low or low to high

With the vehicle stationary and the engine running depress the clutch pedal and move the lever fully forward or fully backwards in **TWO** positive moves to engage the required range. If there is any hesitation in the gear engaging do not force the lever. With the main lever in gear, release the clutch momentarily and then try the transfer lever again.

How to use the transfer gearbox

ADVANCED METHODS

Changing from high to low on the move

When the vehicle is slowing down to a stop and is travelling **NO FASTER THAN 5 km/hr (3 mph)** depress the clutch pedal and push the transfer lever into neutral. Just before the wheels stop turning (clutch still depressed) push the lever fully forward into low (L).

NOTE: Try not to rush the gear change. Use positive and confident moves.

Changing from low to high on the move

Changing from low (L) to high (H) can easily be done without stopping the vehicle. Hold the transfer lever with slight backward pressure on it in preparation for changing. Then, in three simultaneous moves, depress the clutch, release the accelerator and pull the transfer lever into neutral. Release the clutch pedal for approximately 3 seconds, depress it again and move the transfer lever firmly to the high position. Finally, select a suitable main gear for the road speed and release the clutch pedal while depressing the accelerator to take up the drive in the normal way. Continue to drive as normal. This operation can be carried out smoothly and quickly after a little practice. Use firm, positive confident moves. Try to get a rhythm going.

How to use the differential lock

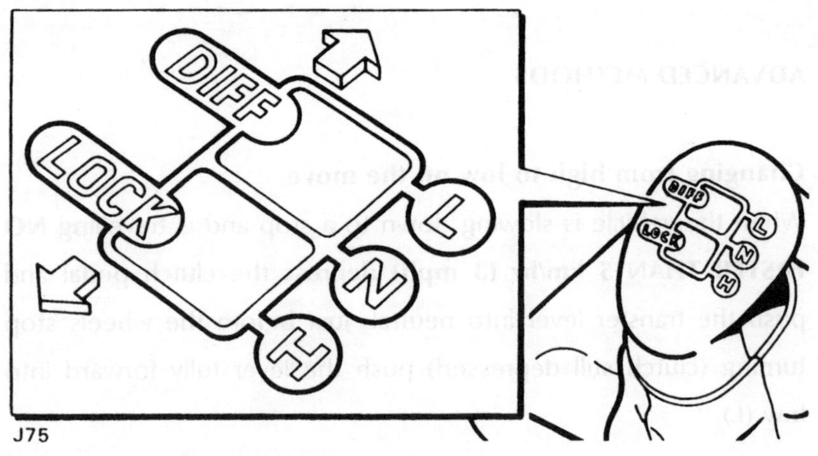

J75

USING THE DIFFERENTIAL LOCK

The **'DIFF LOCK'** can be engaged or dis-engaged at any road speed without depressing the clutch, so long as the vehicle is travelling without wheel slip and in a straight line, or while it is stationary. The differential should only be locked just before slippery or doubtful surface conditions are encountered. Move the transfer gear lever to the left to lock the differential and to the right to unlock it. The differential should always be unlocked as soon as dry firm ground is reached.

CAUTION: DO NOT engage the **'DIFF LOCK'** if one or more wheels are slipping, as this would cause damage to the transmission. If the wheels are slipping, ease off the accelerator before engaging **'DIFF LOCK'**.

How to use the differential lock

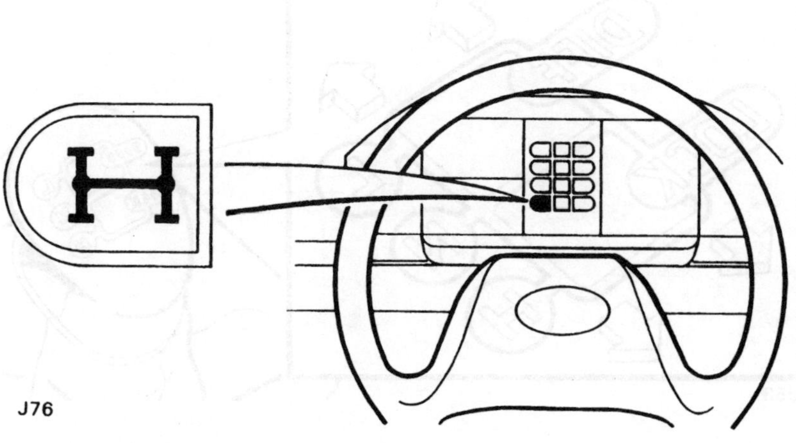

J76

DIFF LOCK WARNING LIGHT

The amber warning light on the instrument panel will be illuminated when the diff lock is engaged. There may be a slight delay between the diff lock being engaged and the light coming on or, between the diff lock being disengaged and the light going off. This is normal.

CAUTION: If the warning light remains on with the diff lock disengaged the transmission is 'wound up'. The vehicle **MUST** be stopped and reversed a short distance to 'unwind' the transmission. The warning light will then be extinguished and the vehicle can proceed as normal. If, after reversing the light remains on, consult your dealer **AS SOON AS POSSIBLE.**

Using the right transfer gear

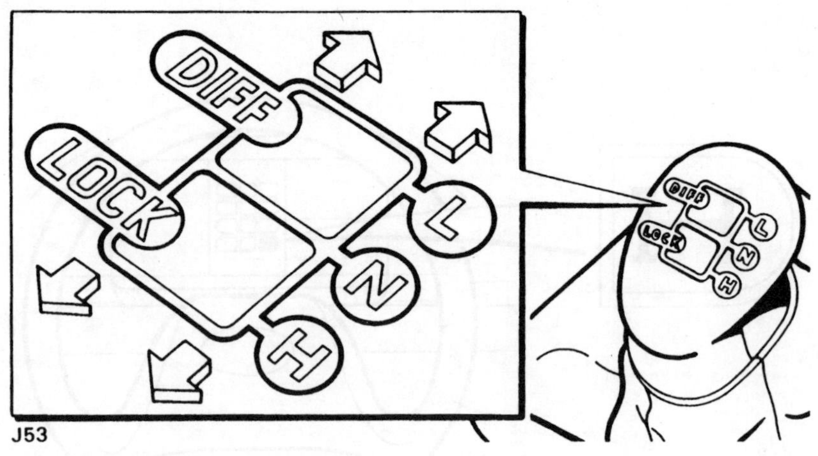

HOW AND WHEN TO USE THE TRANSFER GEARBOX

An introduction to on-road and off-road driving is included on the following pages in the form of an illustrated sequence, which should give new drivers a good idea of how and when to use the transfer gearbox. The gearchange pattern, embossed on the transfer level knob, is repeated in the illustration to show when to select **HIGH, LOW** and **DIFF LOCK.** Fig J53.

More detailed instructions are given on later pages to help improve off-road driving proficiency.

Using the right transfer gear

Driving in town, on a motorway or on wet grass.

Using the right transfer gear

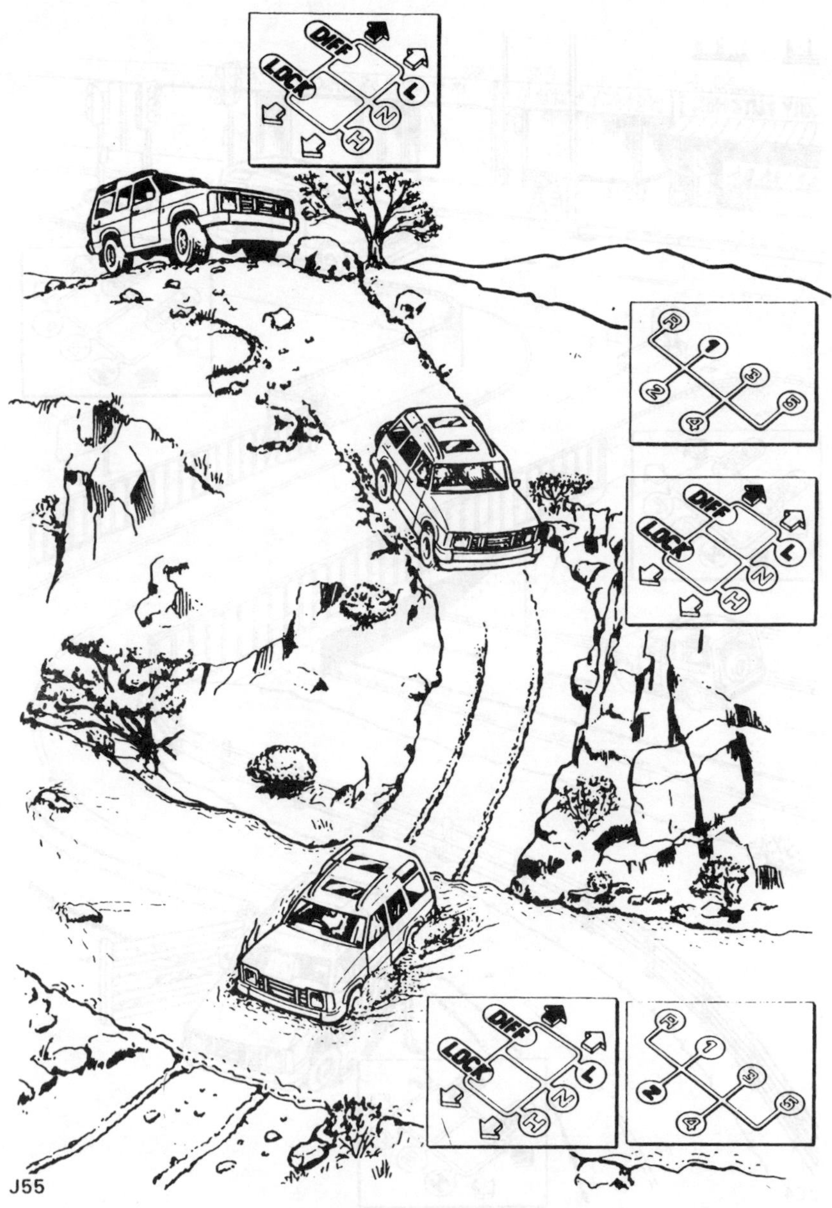

Driving on rocky tracks, down steep slopes and through shallow rivers.

Using the right transfer gear

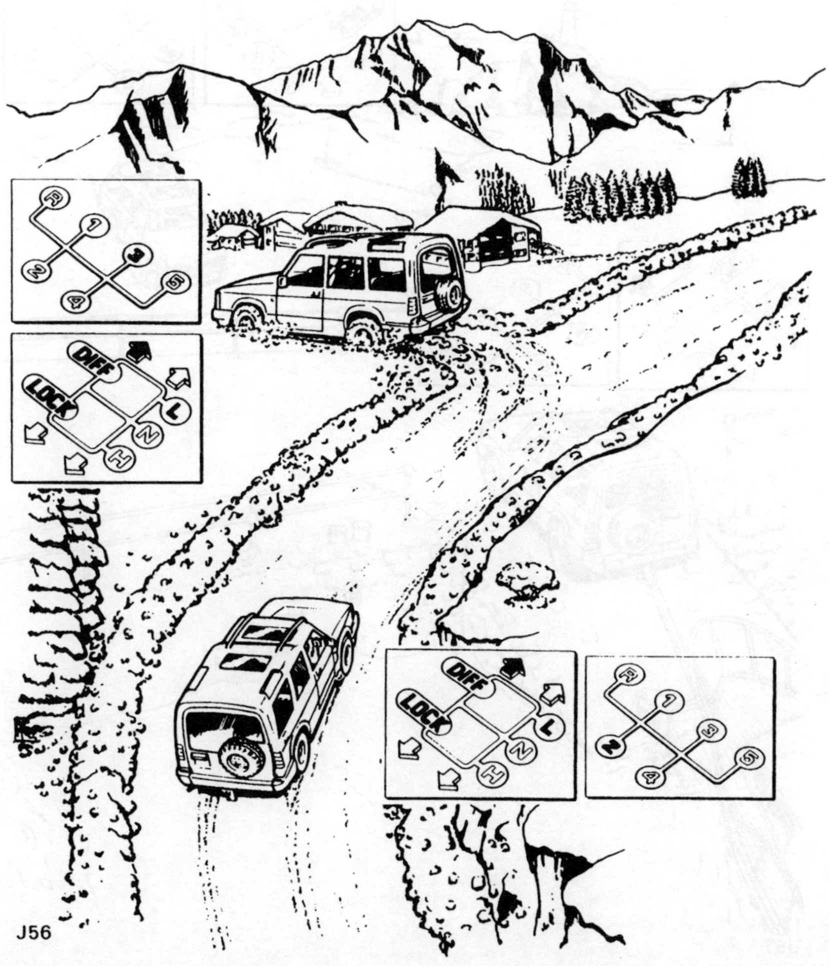

Driving in snow and up steep slopes.

Using the right transfer gear

Driving in soft sand or manoeuving very slowly.

VEHICLE RECOVERY - TOWED

If the vehicle should suffer a breakdown or accident damage and it becomes necessary to make a towed recovery, it is essential to adhere to one of the following procedures depending on the type of tow to be undertaken.

This is because Discovery vehicles have a steering lock.

TOWING THE VEHICLE (ON FOUR-WHEELS)

(a)   Set the main gearbox in neutral.

(b)   Set the transfer box in neutral.

(c)   Turn the starter/steering lock key to the first position to unlock the steering.

(d)   Ensure the differential lock is in the normal 'unlocked' position.

(e)   Secure towing attachment to the vehicle.

(f)   Release the handbrake.

NOTE: Brake servo and power steering assistance, will not be effective unless the engine is running. This will result in a considerable increase in pedal pressure and steering effort being required.

Vehicle recovery

TRANSPORTING THE DISCOVERY ON A TRAILER - Fig. J63

Lashing rings are available on the front and rear chassis members to facilitate the securing of the vehicle to a trailer.

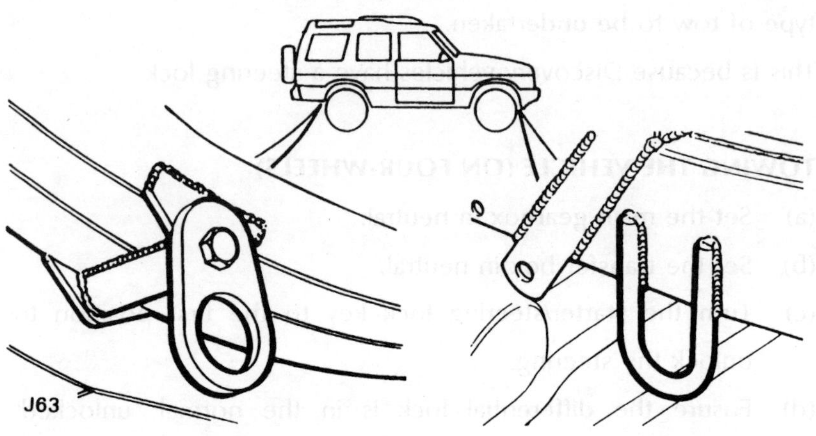

DRIVER'S MAINTENANCE 4

Section 4 includes information for the Driver, on the day-to-day maintenance requirements for the operation of the vehicle. The more comprehensive, 'Routine Maintenance', described in the Service Schedules, should be done by a Land Rover Dealer who will have trained mechanics and full workshop facilities.

Driver's maintenance

The following checks and adjustments should be carried out by the driver or operator, to ensure that the vehicle is ready for daily use.

These tasks are described and illustrated in the following pages and, the recommended lubricants, fluids and quantities are stated in Section 5.

Daily or weekly, depending on operating conditions, and at least every 500 km (250 miles):-

Check/top up engine oil.

Check/top up engine cooling system.

Check/top up windscreen washer reservoir.

Drain fuel sedimenter - Diesel only.

Visually check the brake fluid reservoir. The fluid level must be above the 'MIN' mark. **DO NOT** top up. If the level is low, obtain advice from a Land Rover Dealer.

Check/adjust tyre pressures.

Check tyres for wear or damage.

Check that the handbrake and footbrakes, operate normally.

Check operation of all lights, horn, wipers and washers.

Driver's maintenance

EXTERIOR LAMPS

Owners are under a legal obligation in many countries to maintain all exterior lights in good working order; this also applies to headlamp beam setting, which should be checked at regular intervals by your Dealer.

BATTERY

The battery is fitted in the front of the engine compartment and is a 'Low Maintenance' type that does not require any attention from the driver.

SPARE WHEEL

The spare wheel is mounted on a bracket on the rear door and secured by three nuts.

TOOLS

The lifting jack and wheel chock are stowed in a bay at the front of the engine compartment. The jack handle and tools are stowed in a tool bag under the rear seat.

Driver's maintenance

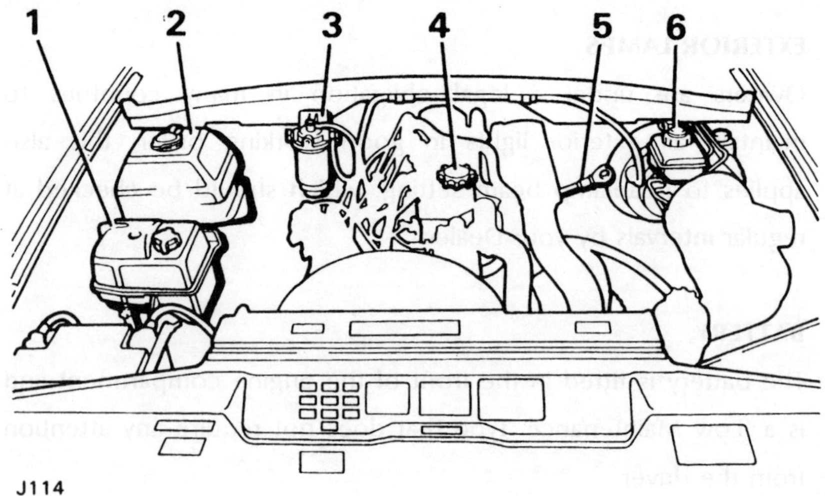

J114

DRIVER'S MAINTENANCE - 200Tdi, LEFT HAND STEERING - Fig. J114

1. Expansion tank - engine coolant.
2. Windscreen washer reservoir.
3. Fuel filter.
4. Engine oil filler cap.
5. Engine oil level dipstick.
6. Brake fluid reservoir.

Instructions on how and when to do the Driver's Maintenance are included on the following pages.

Driver's maintenance

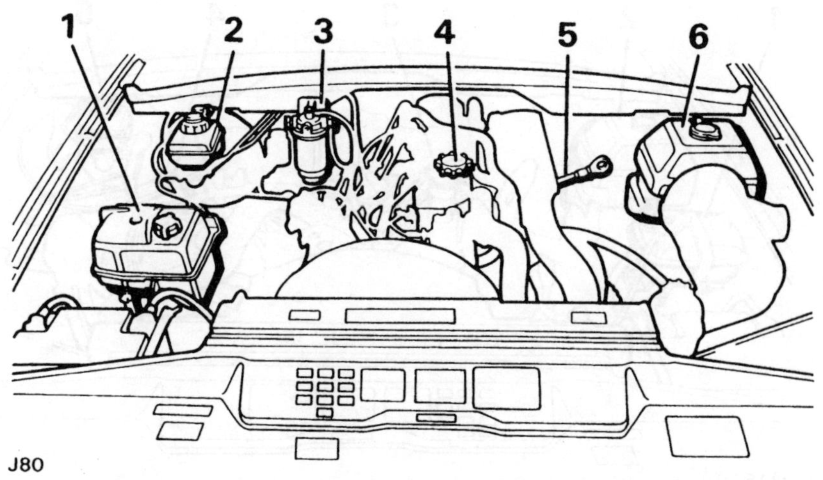

**DRIVER'S MAINTENANCE - 200Tdi, RIGHT HAND STEERING -
Fig. J80**

1. Expansion tank - engine coolant.
2. Brake fluid reservoir.
3. Fuel filter.
4. Engine oil filler cap.
5. Engine oil level dispstick.
6. Windscreen washer reservoir.

Instructions on how and when to do the Driver's Maintenance
are included on the following pages.

Driver's maintenance

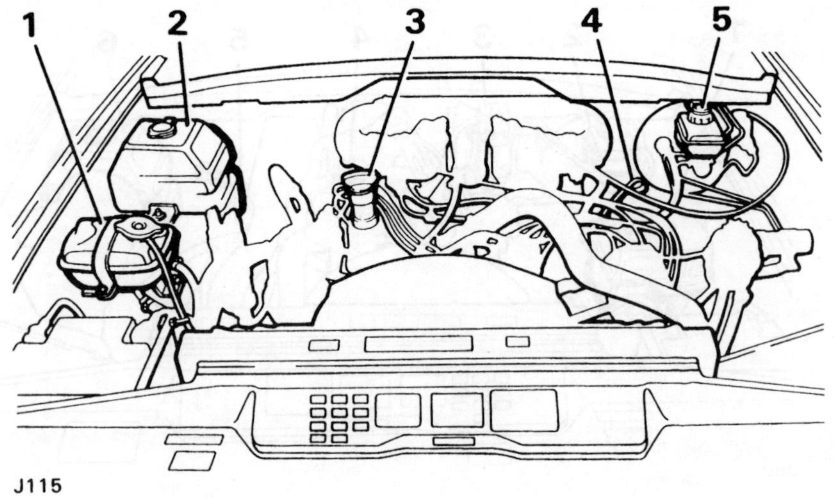

J115

DRIVER'S MAINTENANCE - V8 PETROL, LEFT HAND STEERING - Fig. J115

1. Expansion tank - engine coolant.
2. Windscreen washer reservoir.
3. Engine oil filler cap.
4. Engine oil level dipstick.
5. Brake fluid reservoir.

Instructions on how and when to do the Driver's maintenance are included on the following pages.

Driver's maintenance

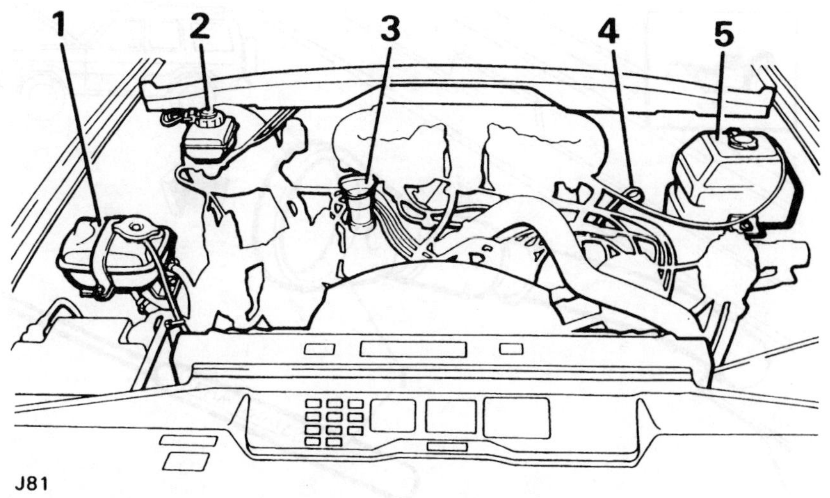

DRIVER'S MAINTENANCE - V8 PETROL, RIGHT HAND

STEERING - Fig. J81

1. Expansion tank - engine coolant.
2. Brake fluid reservoir.
3. Engine oil filler cap.
4. Engine oil level dipstick.
5. Windscreen washer reservoir.

Instructions on how and when to do the Driver's maintenance are included on the following pages.

Diesel engine oil level

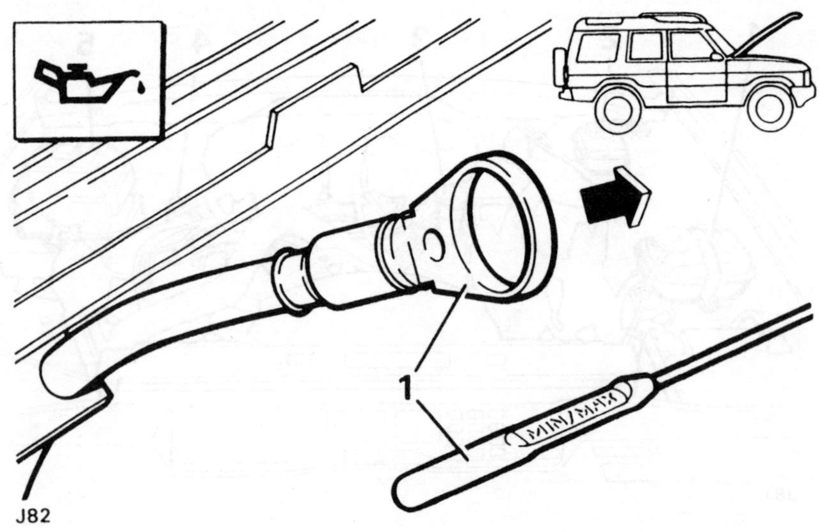

ENGINE OIL LEVEL - 200Tdi ENGINES - Fig. J82 and J108

Check daily or weekly depending on operating conditions and at least every 500 km (250 miles). The oil level should not be allowed to fall below the **MIN** notch on the dipstick (1) located on the left-hand side of the engine. Whenever possible, the oil level should be checked with the engine hot, as follows:

Stand the vehicle on level ground and wait at least five minutes, after the engine has stopped, for the oil to drain back into the engine sump.

Withdraw the dipstick (1) at the left-hand side of the engine, wipe it clean, re-insert it to its full depth and remove a second time to take a reading.

If oil level is between middle and upper notch add no oil.

Diesel engine oil level

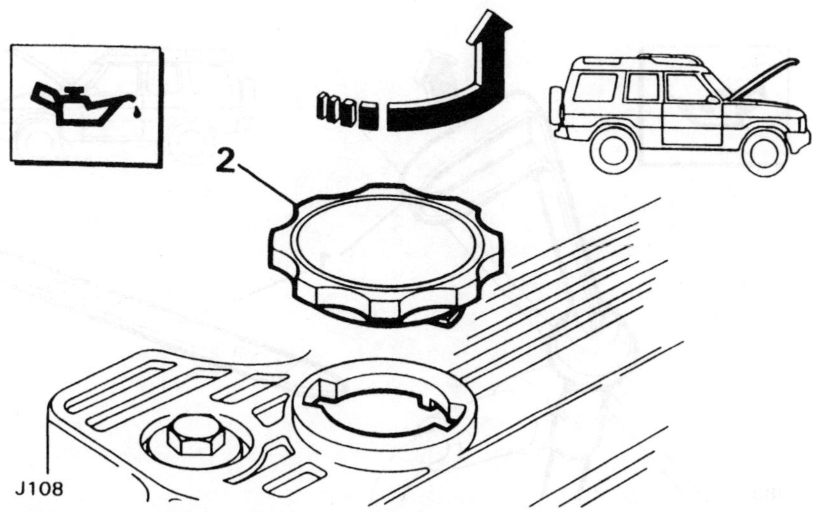

If oil level is between lower and middle notch add one litre only of the correct grade oil through the push-on filler/breather cap (2) on the rocker cover. If the oil level is below the **MIN** notch, add one litre of oil and re-check the level after five minutes. Add further oil as necessary to raise the level between the middle and upper notch. **DO NOT OVERFILL.** See Data Section 5 for recommended engine oils.

IF THE ENGINE IS COLD:

DO NOT start the engine. Ensure that the vehicle is standing on level ground and proceed as above. If it is necessary to re-check oil, or if the engine has been started without being thoroughly warmed up, wait at least 30 minutes to confirm oil level.

CAUTION: Oil level must never be above the **MAX** notch as engine damage may be caused.

Petrol engine oil level

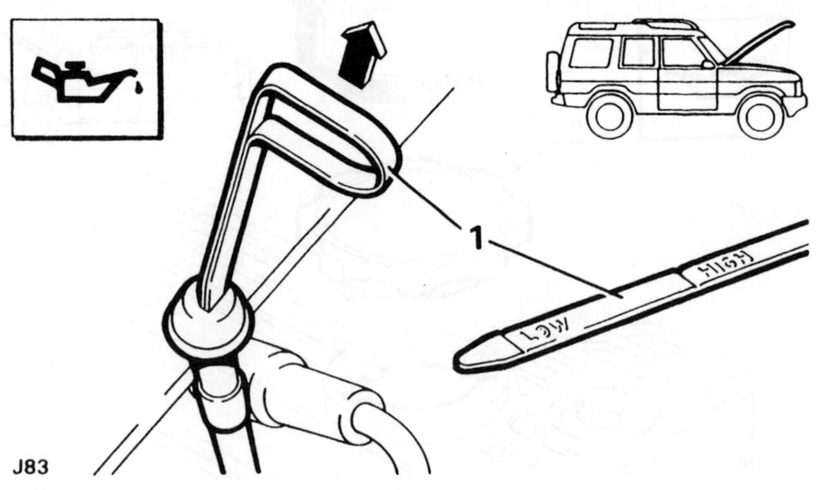

ENGINE OIL LEVEL - V8 PETROL ENGINES - Fig. J83 and J109

Check daily or weekly, depending on operating conditions and at least every 500 km (250 miles).

Whenever possible, the oil level should be checked with the engine hot, as follows:

Stand the vehicle on level ground and wait at least five minutes, after the engine has stopped, for the oil to drain back into the engine sump.

Withdraw the dipstick (1) at the left-hand side of the engine, wipe it clean, re-insert it to its full depth and remove a second time to take a reading. The oil level should not be allowed to fall below the **LOW** mark.

Petrol engine oil level

Add the correct grade of oil, as necessary, through the screw-on filler cap (2) marked **ENGINE OIL** on the right-hand front rocker cover. **Never fill above the HIGH** mark.

See Data Section 5 for recommended engine oils.

IF THE ENGINE IS COLD:

DO NOT start the engine. Stand the vehicle on level ground.

Withdraw the dipstick (1) at the left-hand side of the engine, wipe it clean, re-insert it to its full depth and remove a second time to take a reading. The oil level should not be allowed to fall below the **LOW** mark.

Add the correct grade of oil, as necessary, through the screw-on filler cap (2) marked **ENGINE OIL** on the right-hand front rocker cover. **Never fill above the HIGH** mark.

Engine coolant

ENGINE COOLANT

The coolant level in the expansion tank should be checked daily or weekly depending upon the operating conditions.

200Tdi MODELS

Never run the engine without coolant, not even for a very brief period, otherwise the injectors may be seriously damaged. This is due to the very high rate of heat transfer in the region of the injector nozzles.

ENGINE PROTECTION

To prevent frost damage or corrosion of engine parts it is imperative that the cooling system is filled with a solution of clean water and the correct type of anti-freeze, winter and summer.

In warm climates where frost precautions are not necessary, a solution of clean water and a corrosion inhibitor (Marstons SQ 36) should be used, this is very important because of the aluminium alloy engine parts.

NEVER use salt water, not even with anti-freeze or an inhibitor, otherwise corrosion will occur. In certain countries where the only available water supply may have some salt content, use only clean rainwater or distilled water.

Diesel engine coolant

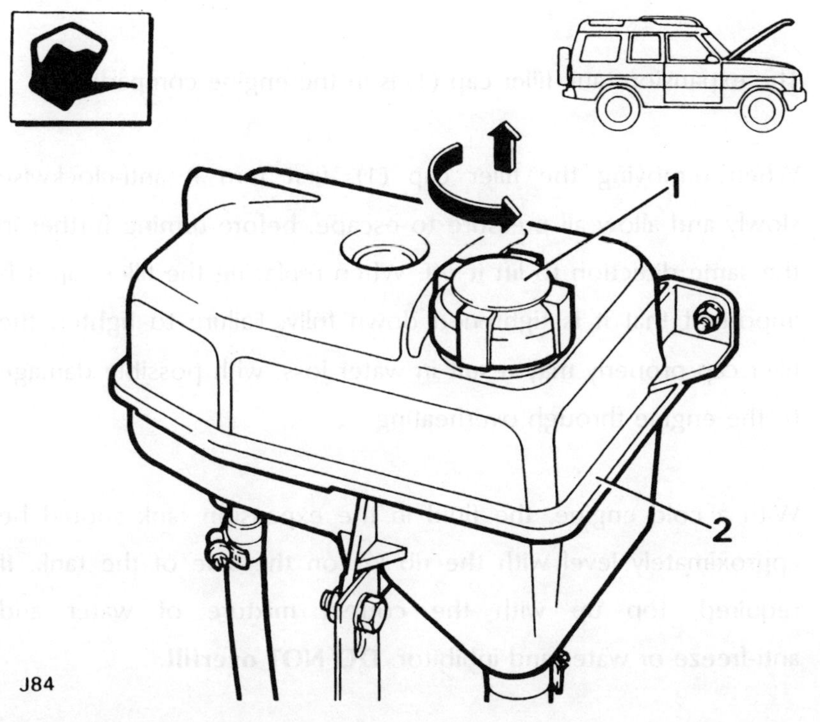

COOLANT LEVEL - 200Tdi ENGINES - Fig. J84

⚠ **WARNING: Do not remove the filler cap or radiator filler plug when the engine is hot because the cooling system is pressurised and personal scalding could result.**

Diesel engine coolant

The expansion tank filler cap (1) is in the engine compartment.

When removing the filler cap (1), first turn it anti-clockwise slowly and allow all pressure to escape, before turning further in the same direction to lift it off. When replacing the filler cap it is important that it is tightened down fully. Failure to tighten the filler cap properly may result in water loss, with possible damage to the engine through overheating.

With a cold engine, the fluid in the expansion tank should be approximately level with the rib (2) on the side of the tank. If required, top up with the correct mixture of water and anti-freeze or water and inhibitor. **DO NOT overfill.**

Petrol engine coolant

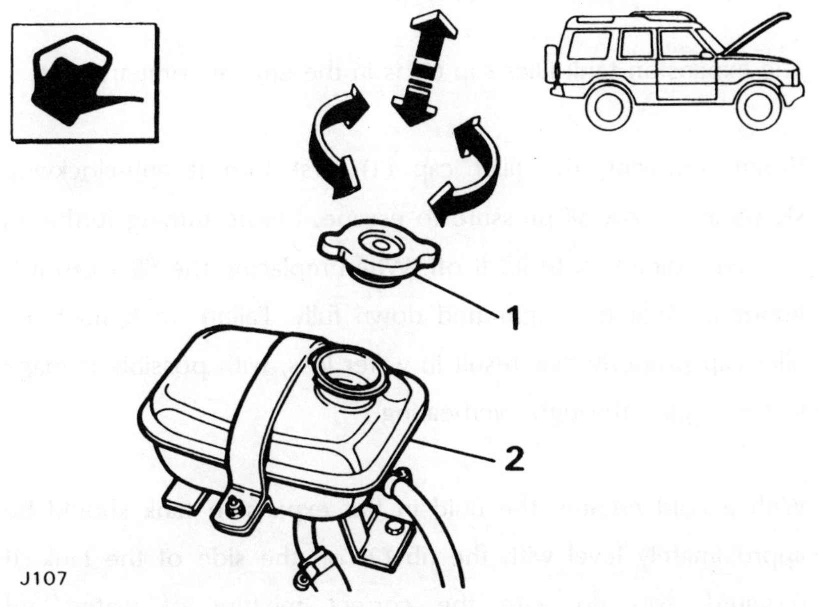

COOLANT LEVEL - V8 PETROL ENGINES - Fig. J107 and J85

⚠ **WARNING: Do not remove the filler cap or radiator filler plug when the engine is hot because the cooling system is pressurised and personal scalding could result.**

Petrol engine coolant

The expansion tank filler cap (1) is in the engine compartment.

When removing the filler cap (1), first turn it anti-clockwise slowly and allow all pressure to escape, before turning further in the same direction to lift it off. When replacing the filler cap it is important that it is tightened down fully. Failure to tighten the filler cap properly may result in water loss, with possible damage to the engine through overheating.

With a cold engine, the fluid in the expansion tank should be approximately level with the rib (2) on the side of the tank. If required, top up with the correct mixture of water and anti-freeze or water and inhibitor. **DO NOT overfill.**

Petrol engine coolant

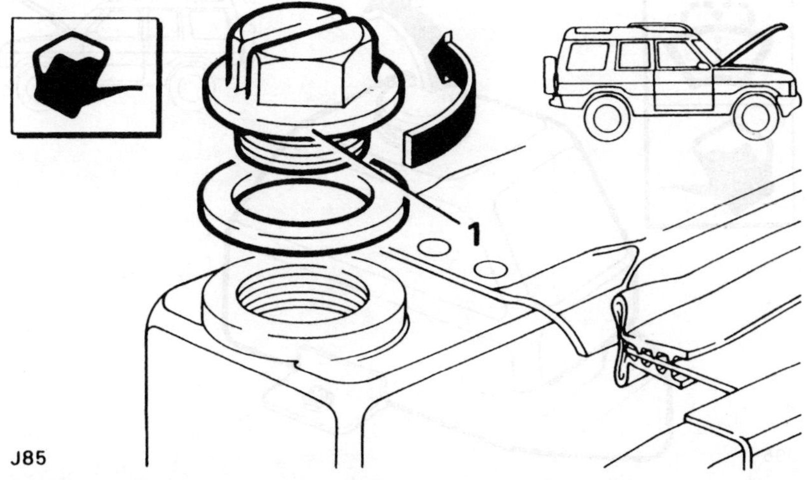

On V8 models, it is important to remove the filler plug in the top of the radiator and check the level of the coolant in the radiator as well as in the expansion tank.

With a cold engine, the fluid in the radiator should be approximately 12 mm (0.5 inch) below the filler neck. If required, top up with the correct mixture of water and anti-freeze or water and inhibitor. **DO NOT overfill.**

When removing the filler plug (1), first turn it anti-clockwise slowly and allow all pressure to escape, before turning further in the same direction to lift it off. When replacing the filler plug it is important that it is tightened down fully. Failure to tighten the filler plug properly may result in water loss, with possible damage to the engine through overheating.

Windscreen washer fluid

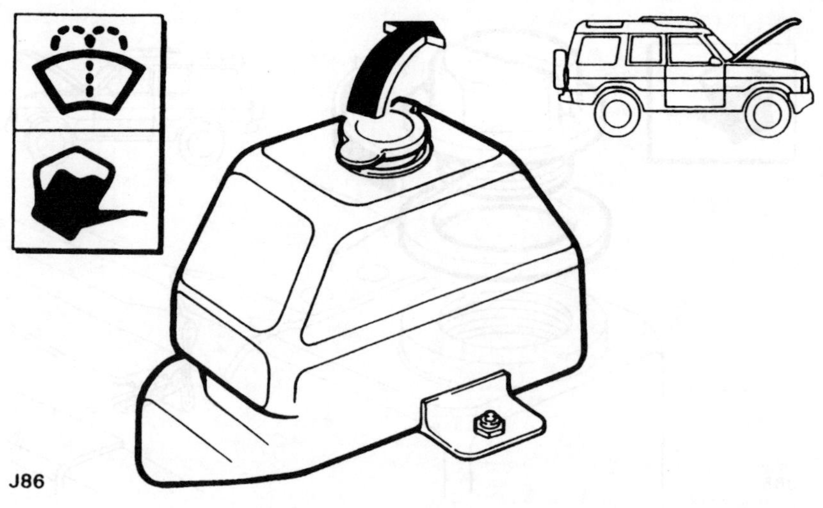

WINDSCREEN AND REAR DOOR WASHER RESERVOIRS - Fig. J86

The windscreen washer reservoir (illustrated), is located in the engine compartment. The reservoir has a large capacity and is fitted with two pumps, one for the front windscreen and one for the rear. An additional pump is fitted, when headlamp washers are specified (option).

Open reservoir cap. Top-up reservoir with clean water to within approximately 25 mm (1 inch) below bottom of filler neck. Use an approved screen washer solvent in the reservoir; this will assist in removing mud, flies and road film. In cold weather, to prevent freezing of the water, add a screen washer solvent containing isopropronal, where this is not available it is permissable to use methylated spirits.

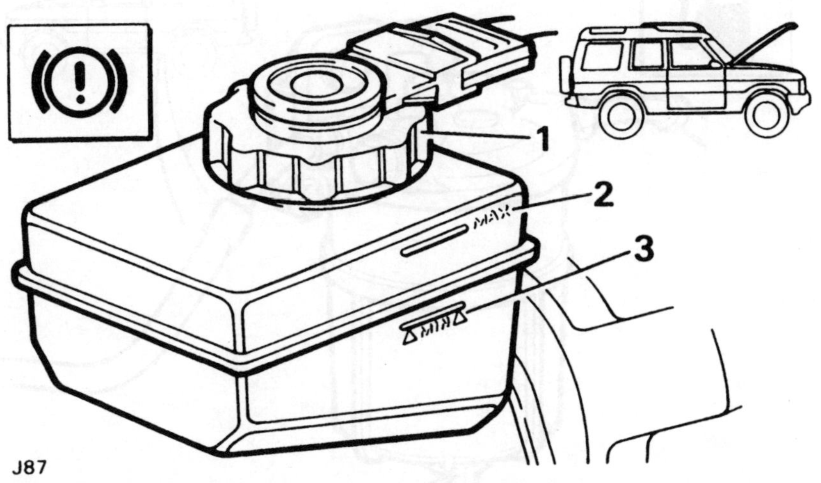

BRAKE FLUID RESERVOIR - Fig. J87

The brake fluid level should be checked visually before driving the vehicle for the first time each day.

DO NOT remove the cap (1). It is important to keep the fluid clean.

With the vehicle standing on level ground, the correct level of the fluid in the reservoir should be between the **MAX** (2) and **MIN** (3) marks.

If the level is **below the MIN mark,** there may be a fault in the braking system. **DO NOT drive the vehicle.**

Contact the nearest garage and have the brakes checked and repaired as necessary.

Diesel fuel sedimenter

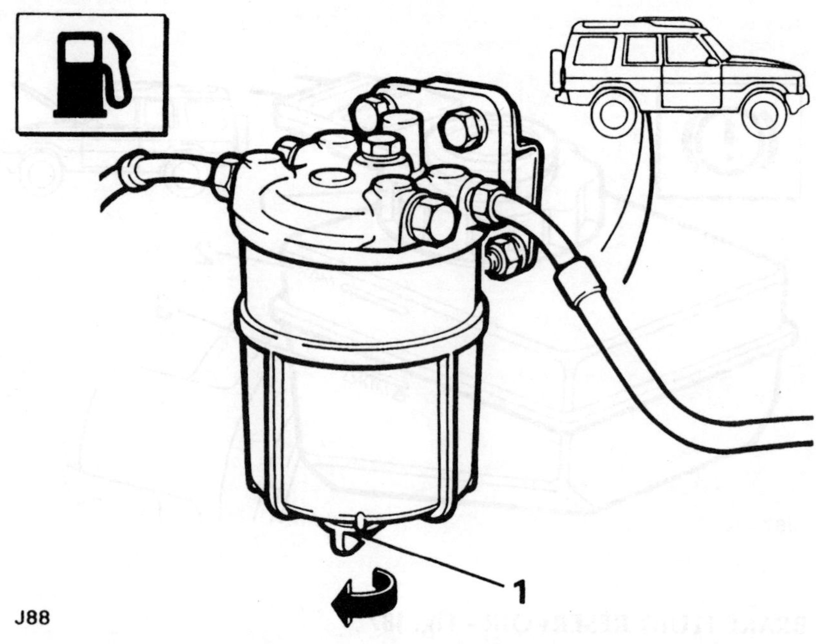

DRAIN FUEL SEDIMENTER - 200Tdi ENGINES - Fig. J88

The sedimenter increases the working life of the fuel filter by removing the larger droplets of water and larger particles of foreign matter from the fuel. The sedimenter is mounted on the chassis side member, near the rear wheel.

DRAIN OFF WATER

Slacken off drain plug (1) to allow water to run out.

When pure diesel fuel is emitted, tighten drain plug. If water is present, the sedimenter should be drained every day until the water content is eliminated and then return to a weekly check.

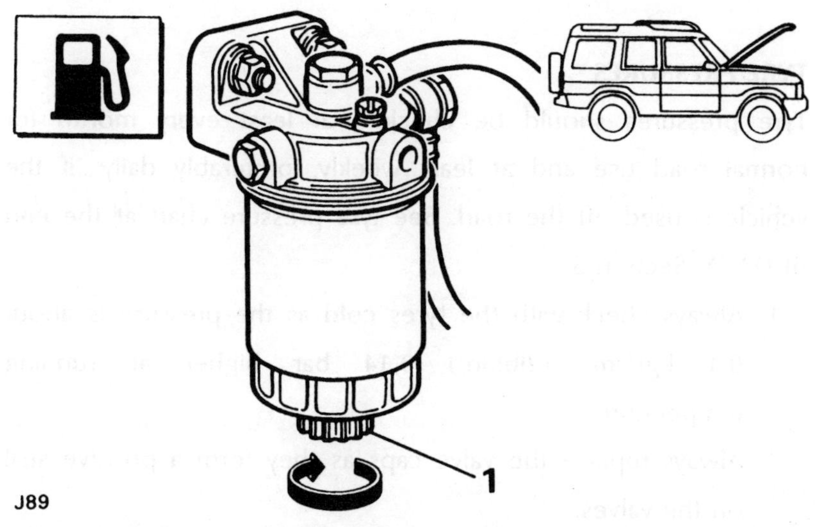

FUEL FILTER, PAPER ELEMENT TYPE (DIESEL MODELS) - Fig. J89

The filter is mounted at the rear of the engine compartment.

Once a month drain off the water as follows:

Slacken off drain plug (1) to allow water to run out.

When pure diesel fuel is emitted, tighten drain plug. If water is present, the filter should be drained every day until the water content is eliminated, and then return to the monthly check.

Tyres and tyre pressures

TYRE PRESSURES

Tyre pressures should be checked at least every month for normal road use and at least weekly, preferably daily, if the vehicle is used off the road. See tyre pressure chart at the end of DATA, Section 5.

1. Always check with the tyres cold as the pressure is about 0,1 kgf/cm² (2lbf/in²) 0,14 bar higher at running temperature.

2. Always replace the valve caps as they form a positive seal on the valves.

3. Any unusual pressure loss in excess of 0,05 to 0,20 kgf/cm² (1 to 3 lbf/in²) 0,07 to 0,21 bar per week should be investigated and corrected.

4. Always check the spare wheel so that it is ready for use at any time.

5. Maximum tyre life and performance will only be obtained if the tyres are maintained at the correct pressure.

Check tyres for tread depth and visually for external cuts in the fabric, exposure of ply or cord structure

The tread should be measured at every maintenance inspection and when the tread has worn to a remaining depth of 1,6 mm (1/16 in), new tyres should be fitted. Do not continue to use tyres that have worn to the recommended limit or the safety of

Tyres and tyre pressures

the vehicle could be affected and legal regulations governing tread depth may be broken. At the same time remove embedded flints etc. from the tyre treads with the aid of a penknife or similar tool and check that the tyres have no breaks in the fabric or cuts to sidewalls etc. Clean off any oil or grease on the tyres using white spirit sparingly. Check that there are no lumps or bulges in the tyres or exposure of the ply or cord structure.

It is illegal in the UK and many other countries to continue to use tyres with excessively worn tread. Tyre wear should be checked at every maintenance inspection.

Jack and tools

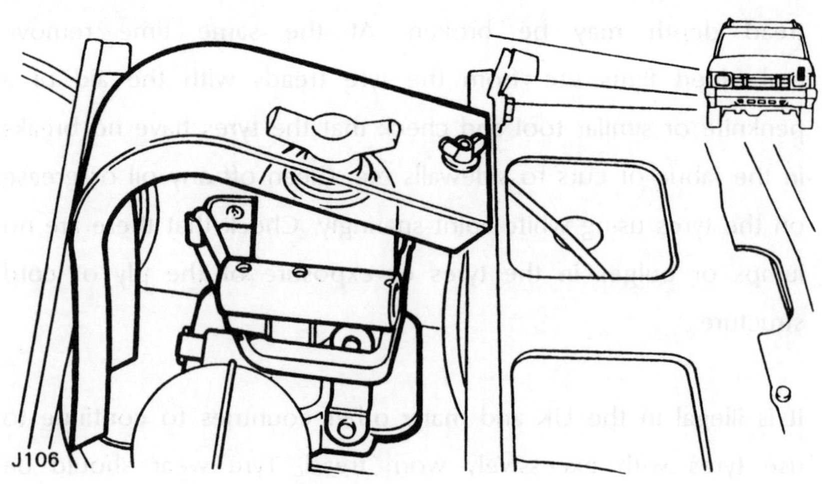

JACK AND WHEEL CHOCK - Fig. J106

The jack and wheel chock are stowed in a bay at the front of the engine compartment. The jack is retained by a rubber strap and the chock is secured by a wing nut.

⚠ **WARNING: The jack and wheel chock, and other items under the bonnet, will be hot if the engine has recently been running.**

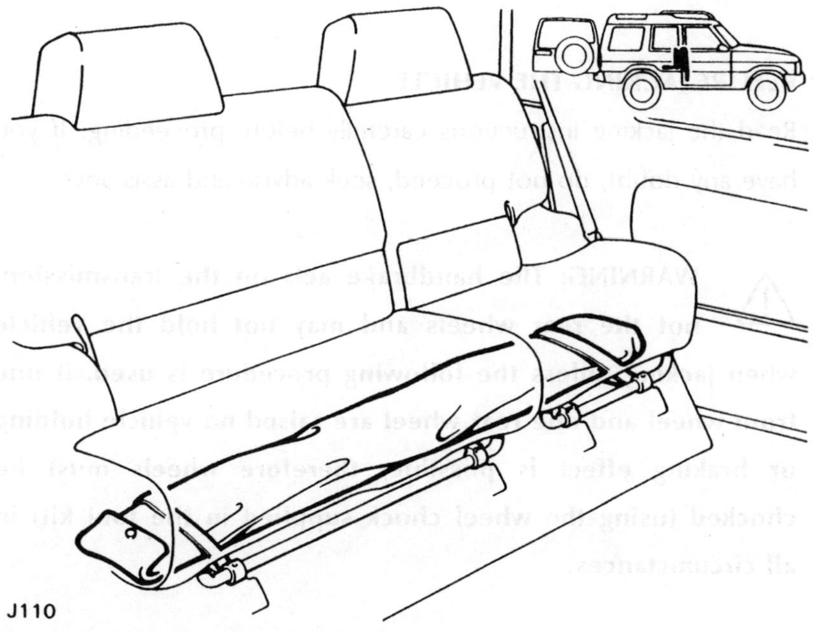

J110

JACK HANDLE AND TOOLS - Fig. J110

The jack handle and tools are stowed in a tool bag under the rear seat, and secured in position with straps.

Jacking

BEFORE JACKING THE VEHICLE

Read the jacking instructions carefully before proceeding. If you have any doubt, do not proceed, seek advise and assistance.

⚠️ **WARNING: The handbrake acts on the transmission, not the rear wheels and may not hold the vehicle when jacking unless the following procedure is used. If one front wheel and one rear wheel are raised no vehicle holding or braking effect is possible, therefore wheels must be chocked (using the wheel chock supplied in the tool kit) in all circumstances.**

The jack should be used on level and firm ground. Always engage the differential lock before jacking. The differential lock is only engaged if the warning light is illuminated with the ignition/starter switch switched on. No person should remain in a vehicle being jacked. Apply the handbrake. Engage first gear in the main gearbox. Engage low gear in the transfer box. Turn off the ignition/starter switch and remove the key.

⚠️ **WARNING: If the vehicle is coupled to a trailer, disconnect the trailer from the vehicle before commencing jacking. This is to prevent the trailer pulling the vehicle off the jack and causing personal injury.**

Jack and wheel chock

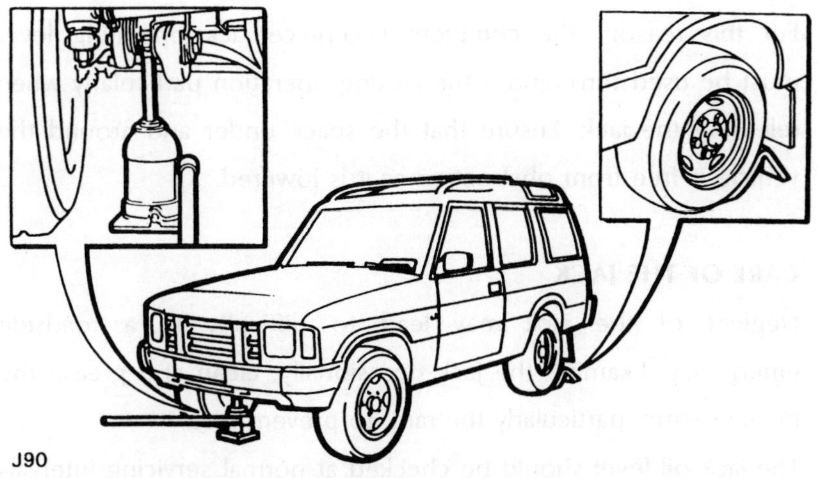

J90

USING THE WHEEL CHOCK - Fig. J90

Before jacking up a front wheel, the wheel chock should be positioned behind the rear wheel diagonally opposite the wheel to be raised.

Before jacking up a rear wheel, the chock should be positioned in front of the front wheel diagonally opposite the wheel to be raised.

⚠️ **WARNING: It is unsafe to work under the vehicle using only the jack to support it. Always use heavy duty stands or other suitable supports to provide adequate safety. Care must be taken to avoid accidental contact with any underbody parts but especially the hot exhaust system components likely to cause personal injury during raising or lowering of the vehicle.**

Jack and wheel chock

For this reason, the complete two-piece jack operating lever must be used throughout the jacking operation particularly when releasing the jack. Ensure that the space under and around the vehicle is free from obstruction as it is lowered.

CARE OF THE JACK

Neglect of the jack may lead to difficulty in a roadside emergency. Examine the jack occasionally, clean and grease the moving parts, particularly the ram, to prevent rust.

The jack oil level should be checked at normal servicing intervals and if necessary topped up with a hydraulic oil with a viscosity to BS4231 grade 32 and ISO proof 32.

To avoid contamination, the jack should be always retrurned to its fully closed position before stowage in its correct position. Always firmly tighten the wing nut to secure the wheel chock and jack.

Jack and jack handle

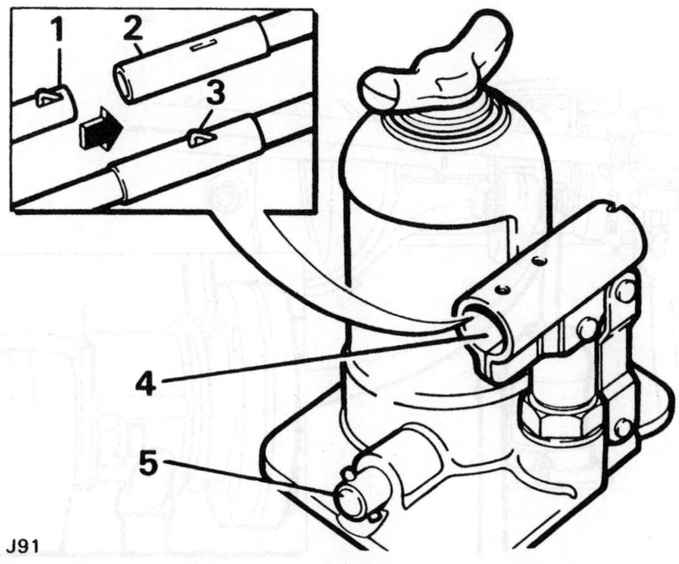

ASSEMBLING THE JACK OPERATING LEVER - Fig. J91

Assemble the two-piece jack operting lever as shown by inserting piece 1 into piece 2 ensuring that when assembled the spring clip 3 protrudes from the engagement slot.

OPERATING THE HYDRAULIC JACK

Check that the release valve (5) at the bottom of the jack body is closed (turned fully clockwise). With the operating lever inserted into the socket (4), pump the lever up and down to raise the jack.

To lower the jack, withdraw the lever from the socket, engage it over the pegs on the release valve (5) and use it to slowly turn the release valve anti-clockwise. Downwards pressure on the cradle will lower it.

Jacking front wheels

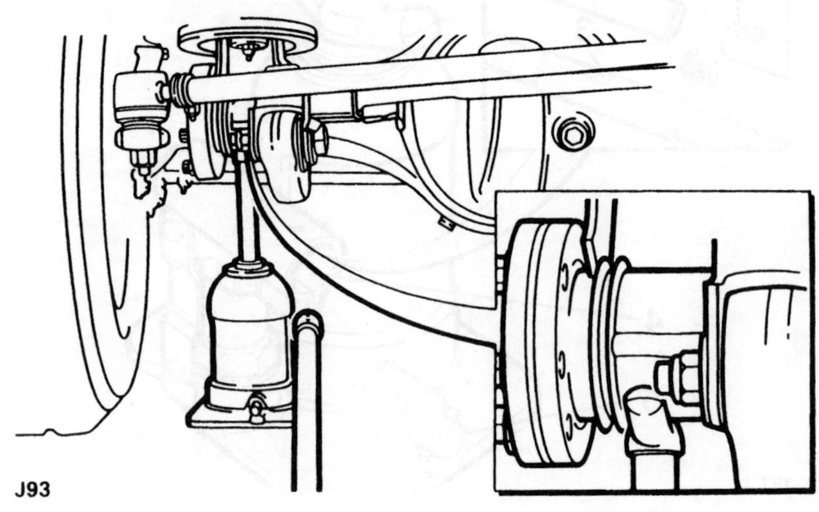

POSITIONING THE JACK TO RAISE A FRONT WHEEL - Fig. J93

Enter the jack underneath the vehicle from directly in front of the jacking point.

CAUTION: Never attempt to use the jack from the side of the vehicle.

Position the jack so that, when raised, it will engage with the front axle casing immediately below the coil spring. The cradle of the jack must locate between the flange at the end of the axle casing and a large bracket to which front suspension members are mounted.

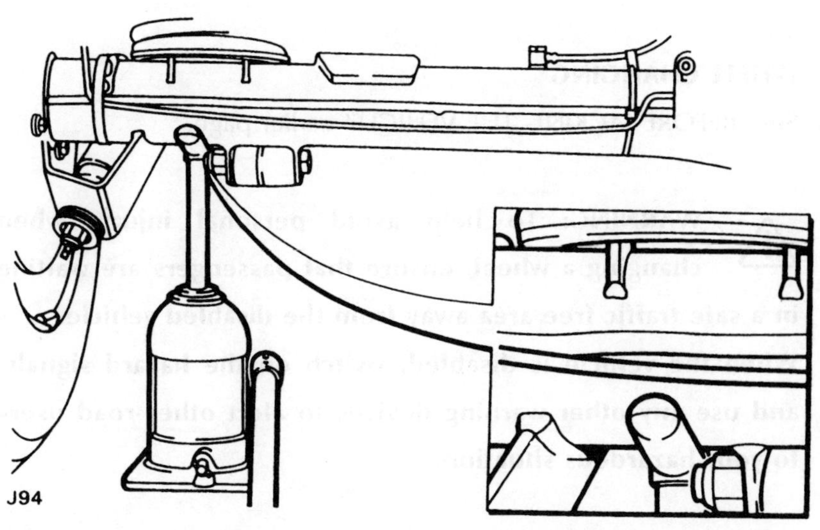

POSITIONING THE JACK TO RAISE A REAR WHEEL - Fig. J94

Push the rear mudflap up behind the tyre to allow a clear view of the jacking point. Return the flap to its correct position at completion of jacking.

Enter the jack underneath the vehicle from the rear and directly in line with the jacking point.

Position the jack so that when it is raised, it will engage with the rear axle casing immediately below the coil spring and as close as possible to the shock absorber mounting bracket.

Wheel changing

WHEEL CHANGING

See 'BEFORE JACKING THE VEHICLE' earlier page.

⚠ **WARNING: To help avoid personal injury, when changing a wheel, ensure that passengers are waiting in a safe traffic free area away from the disabled vehicle. When the vehicle is disabled, switch on the hazard signals, and use any other warning devices to alert other road users to your hazardous situation.**

Using the wheel nut wrench supplied in the tool kit, initially slacken the nuts on the wheel to be removed before jacking the vehicle. Place the wheel chock in position and jack up the corner of the vehicle. When the wheel is clear of the ground, remove wheel nuts and lift off wheel. If available, place a drop of oil or grease on the wheel studs to assist in replacement.

⚠ **WARNING: DO NOT let the spare wheel drop from its mounting bracket on the rear door, as it is heavy and could cause personal injury. If necessary, get help to lift the wheel off and onto the rear door bracket.**

Wheel changing

Fit spare wheel and tighten the nuts loosely.

Lower the vehicle to the ground and finally tighten the nuts. When using the wheelbrace from the vehicle tool kit apply hand pressure only.

DO NOT use foot pressure or extension tubes as this could overstress the wheel studs.

Remember to disengage the differential lock after road wheel has been replaced.

⚠️ **WARNING: Always secure tools, jack and spare wheel in their proper storage positions after wheel changing.**

ROAD WHEEL NUTS

Check road wheel nuts for tightness, torque 128,8 Nm ± 6,8 Nm (95 ± 5 lbf ft). **DO NOT** overtighten.

Diesel engine wading plug

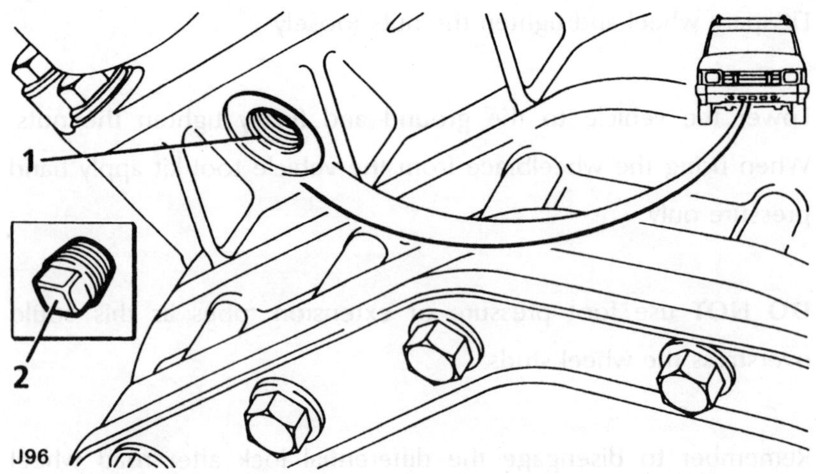

WADING PLUG FOR ENGINE FRONT TIMING COVER - 200Tdi MODELS - Fig. J96

⚠ **WARNING: DO NOT work underneath the vehicle unless it is safely parked and the wheels chocked, or it is supported by heavy duty stands, otherwise the vehicle could move causing personal injury.**

The timing cover can be completely sealed to exclude mud and water under severe wading conditions by fitting a plug in the drain hole (1) at the bottom of the cover.

A suitable plug is available from your Land Rover Dealer's Parts Department.

Diesel engine wading plug

The plug (2) should only be fitted when the vehicle is expected to undertake wading or very muddy work. When the plug is in use it must be removed periodically to allow any oil to drain off before the plug is replaced.

NOTE: There should not be any oil in the timing cover, but if there is, the cause should be investigated as soon as possible, as the timing belt will deteriorate if it becomes contaminated with oil. When the plug is not in use it should be stowed in the tool kit.

Petrol engine wading plug

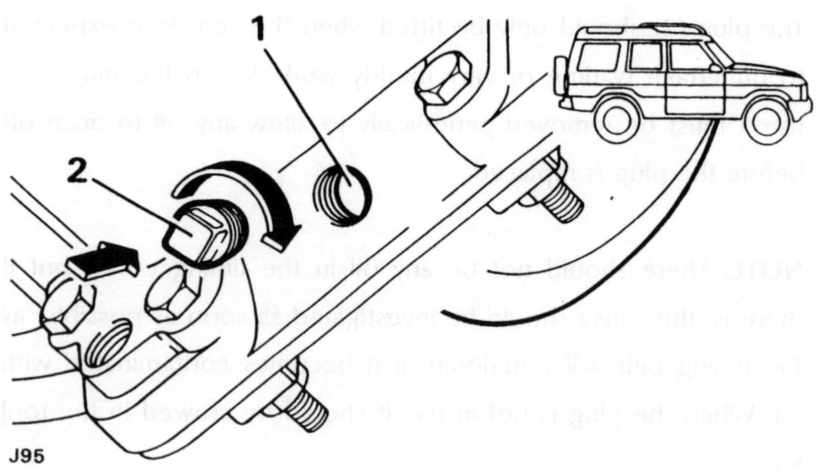

WADING PLUG FOR FLYWHEEL HOUSING - V8 PETROL MODELS ONLY - Fig. J95

⚠️ **WARNING: DO NOT work underneath the vehicle unless it is safely parked and the wheels are chocked, or it is supported by heavy duty stands, otherwise the vehicle could move causing personal injury.**

If the vehicle is to be used for wading, a sealing plug should be fitted to the drain hole in the flywheel housing to prevent the entry of water and mud. A suitable plug is available from your Land Rover Dealers Parts Department.

Petrol engine wading plug

The flywheel housing can be completely sealed to exclude mud and water under severe wading conditions, by fitting a plug in the drain hole (1) at the bottom of the housing. The plug (2) should only be fitted when the vehicle is expected to undertake wading or very muddy work. When the plug is in use it must be removed periodically and all oil allowed to drain off before the plug is replaced. When the plug is not in use it should be stowed in the vehicle tool kit.

Fuse box

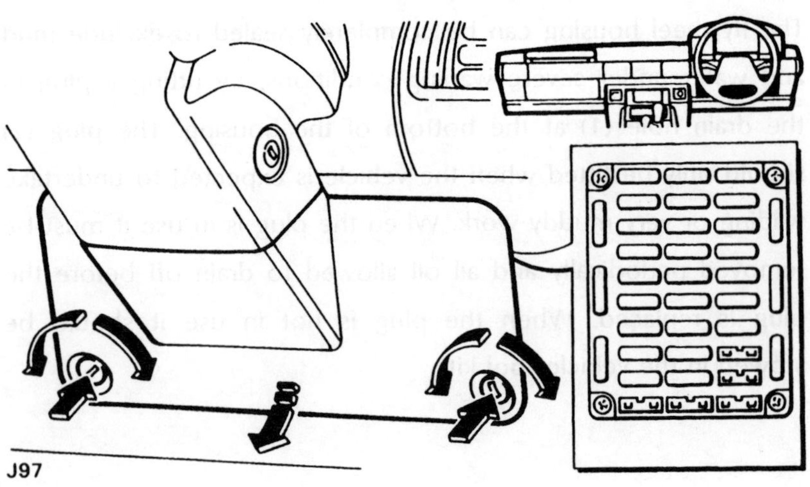

J97

FUSE BOX - Fig. J97

The fusebox is fitted behind the fascia below the steering wheel and can be reached by using a coin to turn the two screws and withdrawing the access panel.

The fuses are colour coded with their amp rating, as follows;

VIOLET	3	BLUE	15
TAN	5	YELLOW	20
BROWN	7.5	WHITE	25
RED	10	GREEN	30

A label in the fuse box cover shows the circuits protected, the fuse colours and their fitted position.

NOTE: Early production vehicles have two additional, 7.5 amp in line fuses. One is in the white lead close to the switch for the Heated Rear Window and is only required if electrically operated mirrors are fitted. The other fuse is in the green lead to the gearbox reverse light switch and is fitted close to the bulkhead connector at the rear of the engine compartment.

Changing headlamp bulb

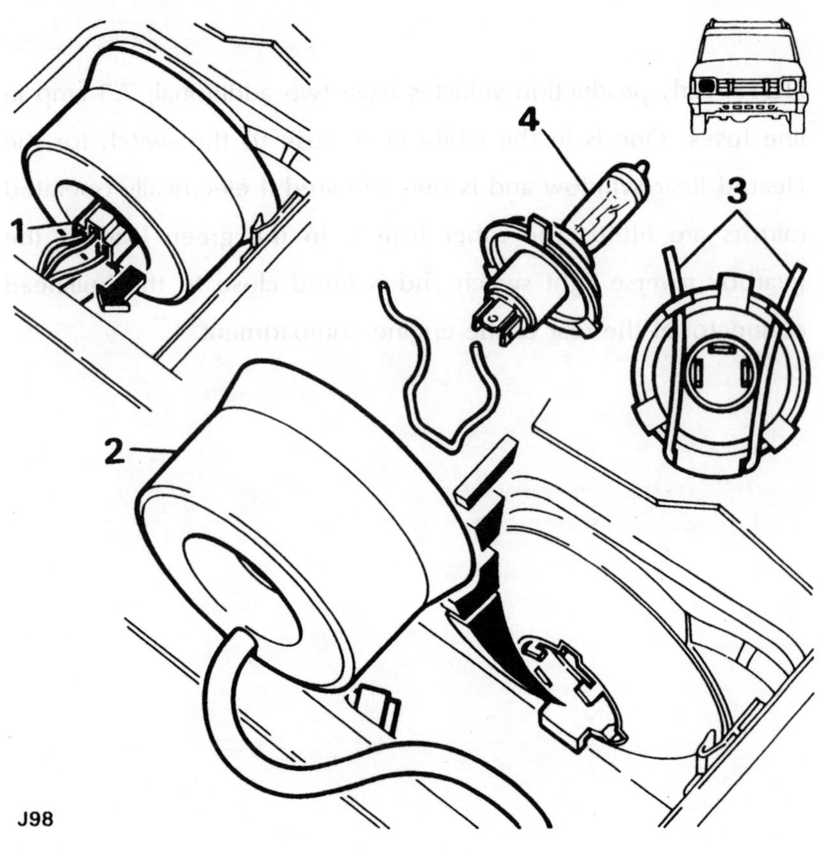

J98

HEADLAMP BULB - Fig. J98

Pull off the connector (1) and the headlamp cover (2). Squeeze the ends of the spring clip (3) inward and unhook the clip at the bottom. Change the bulb (4) and reassemble, noting that the bulb (4) and connector (1) will only fit in one position. Make sure that the spring clip (3) is hooked at the bottom and engaged in the slots at the top.

Changing side lamp bulb

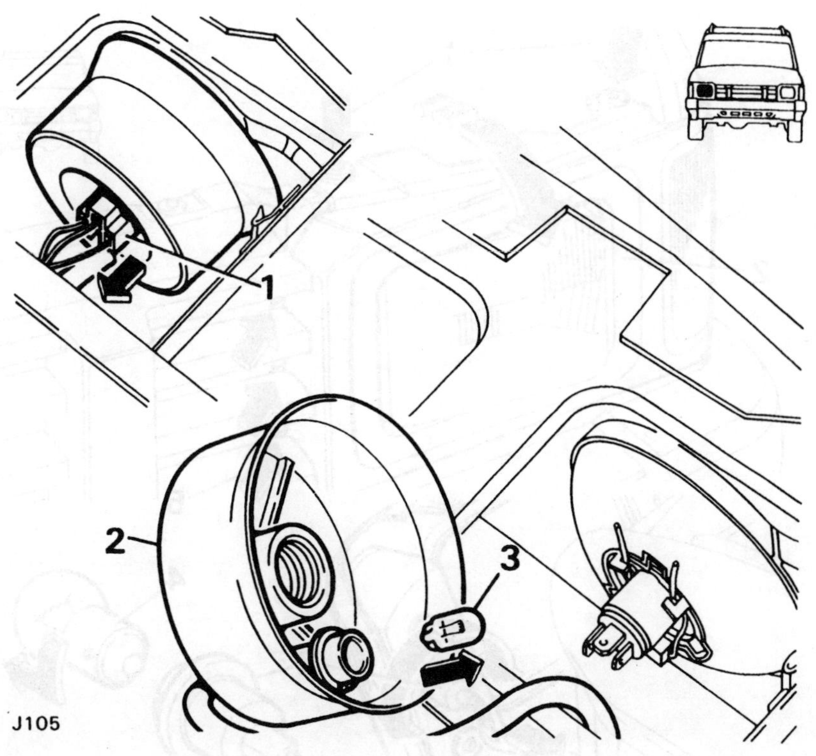

J105

FRONT SIDE LAMP BULB - Fig. J105

Pull off the connector (1) and the headlamp cover (2). Change the bulb (3) and reassemble, noting that the connector (1) will only fit in one position.

Changing front indicator bulb

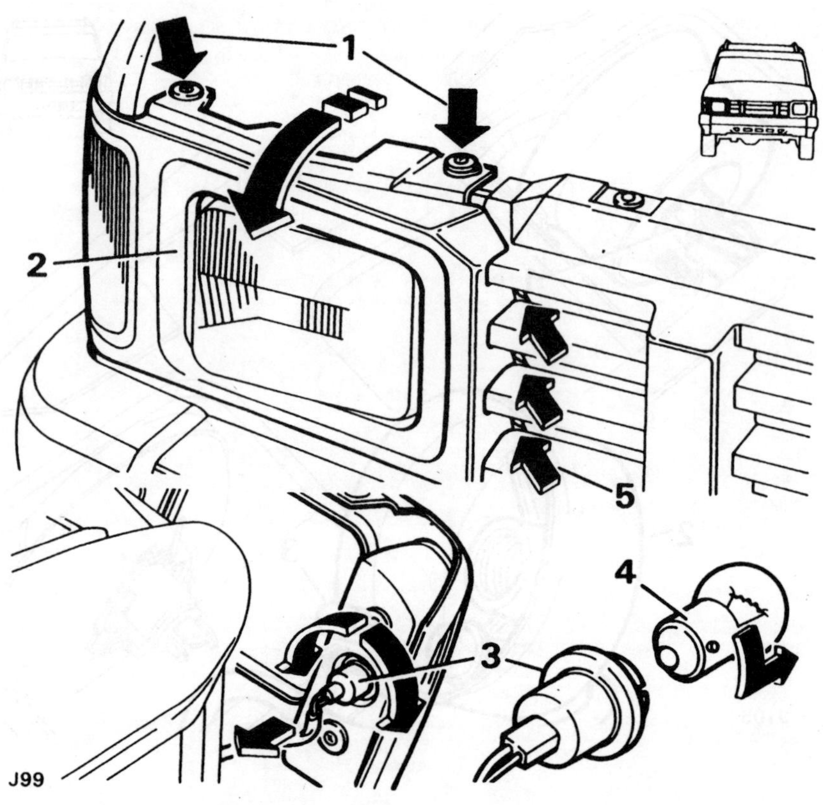

FRONT INDICATOR LAMP - Fig. J99

Remove the screws (1) and pull the frame (2) forward. Withdraw the bulb holder (3) and change the bulb (4). Reassemble, making sure the lugs (5) are engaged. Take care to ensure that the frame is correctly aligned with the wing during reassembly.

Changing side repeater bulb

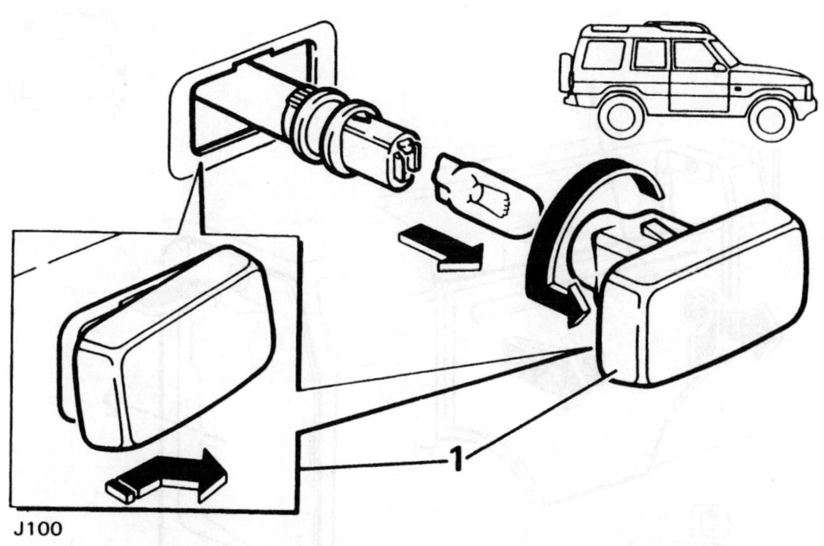

J100

SIDE REPEATER LAMP BULB - Fig. J100

Push the lens (1) forward while lifting the rear edge, then pull the lamp out and remove the lens. Change the bulb and reassemble, noting that the lens will only fit one way.

Changing rear lamp bulbs

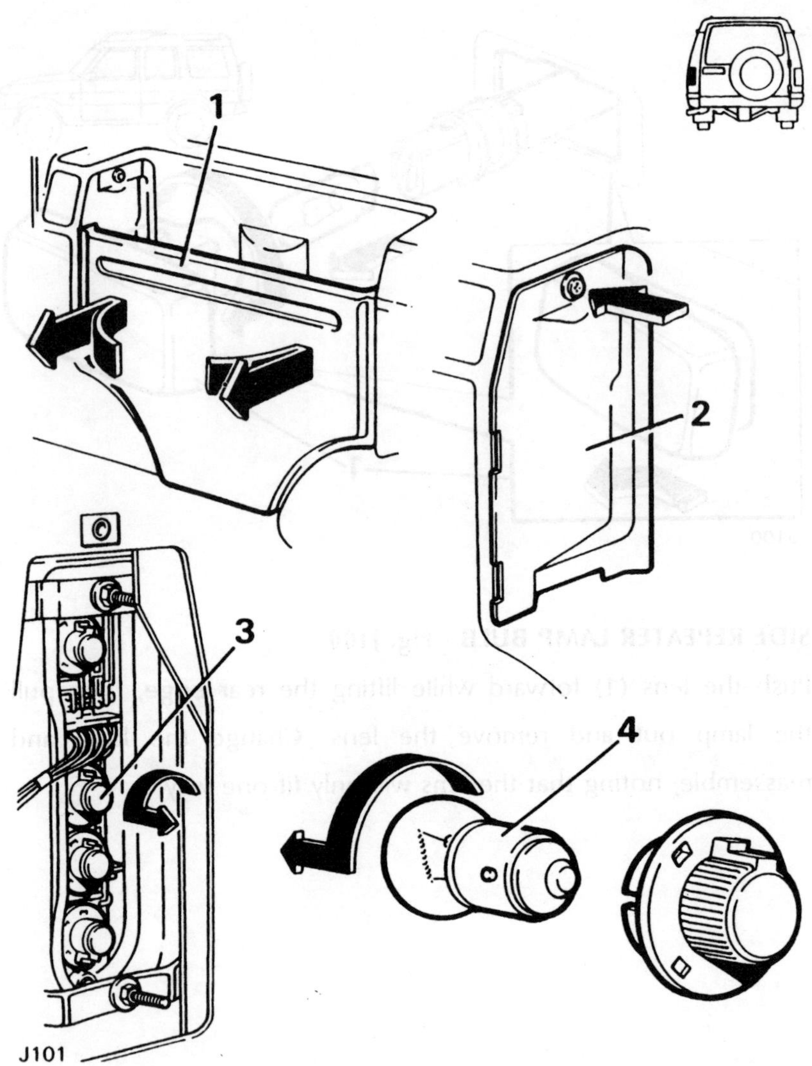

REAR LAMP BULBS - Fig. J101

Pull out the side seat or pocket cover (1) as applicable. Remove the access panel (2), twist out the bulb holder (3), change the bulb (4) and reassemble.

Changing number plate bulb

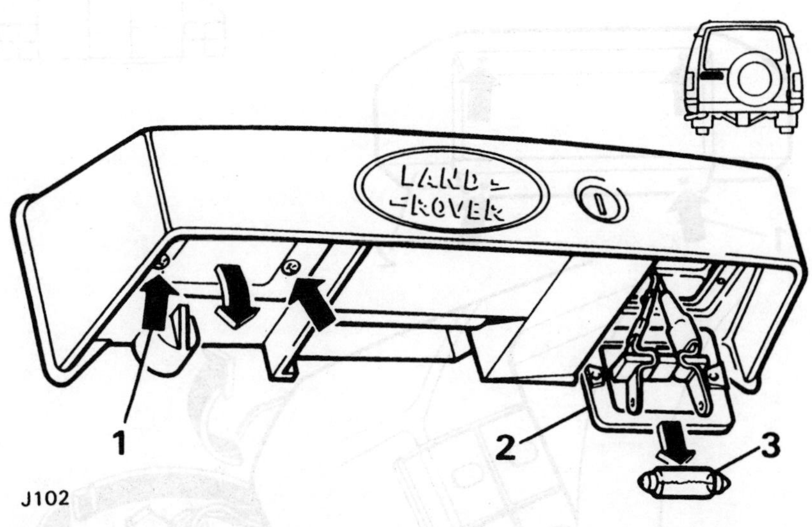

REAR NUMBER PLATE BULB - Fig. J102

Remove the screws (1) and pull the lens (2) down. Change the bulb (3) and reassemble.

Changing panel switch bulbs

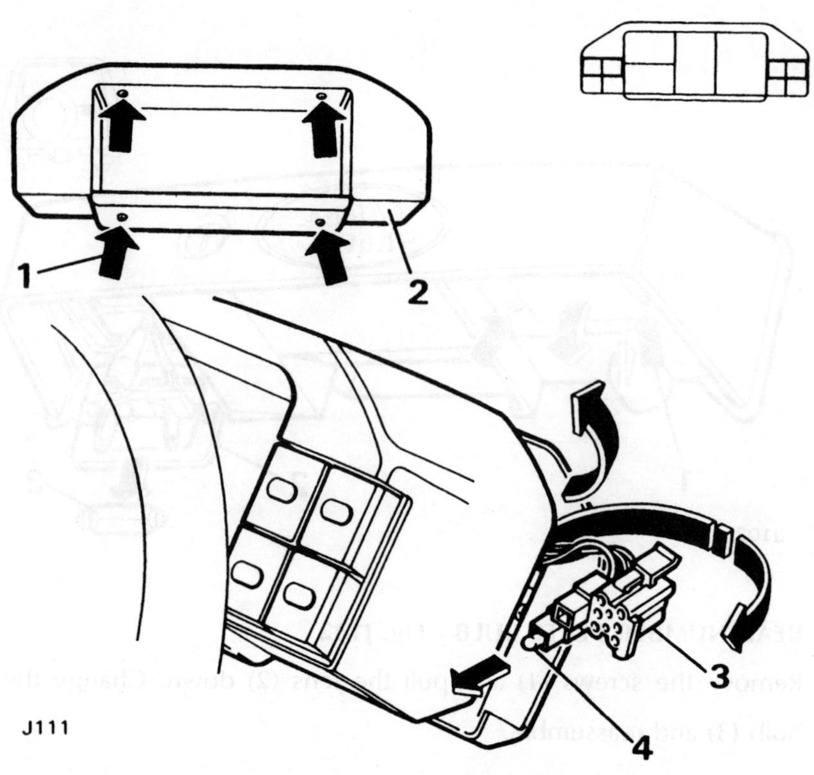

PANEL SWITCH ILLUMINATION - Fig. J111

Remove the four screws (1) and pull the instrument cover (2) forward. Pull the bulb holder (3) from the back of the switch panel, change the bulb (4) and reassemble.

Changing interior lamp bulbs

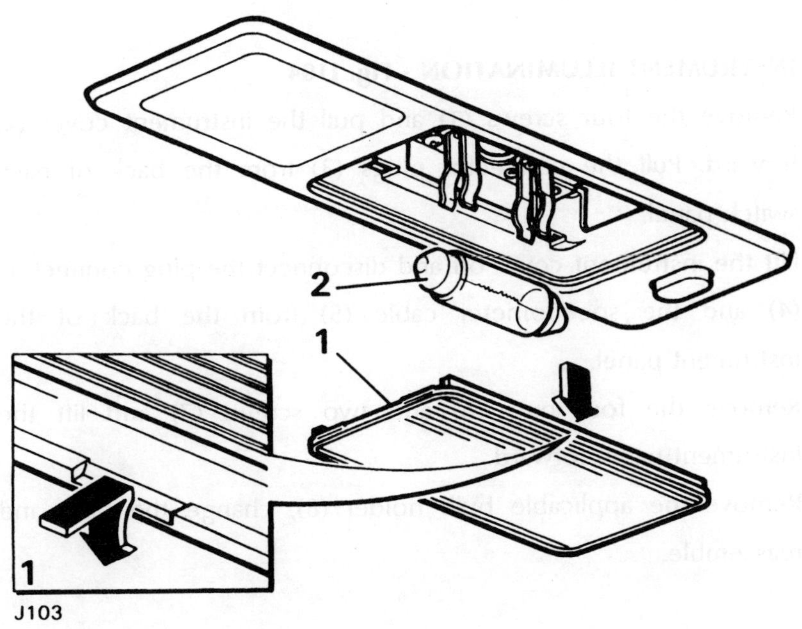

INTERIOR LAMP BULBS - Fig. J103

Use a small screwdriver to prise the lens (1) from its housing.

Change the bulb (2) and reassemble.

Changing instrument bulbs

INSTRUMENT ILLUMINATION - Fig. J104

Remove the four screws (1) and pull the instrument cover (2) forward. Pull the connector plugs (3) from the back of each switch panel.

Lift the instrument cover off and disconnect the plug connectors (4) and the speedometer cable (5) from the back of the instrument panel.

Remove the four nuts (6) and two screws (7) and lift the instrument panel forward.

Remove the applicable bulb holder (8), change the bulb and reassemble.

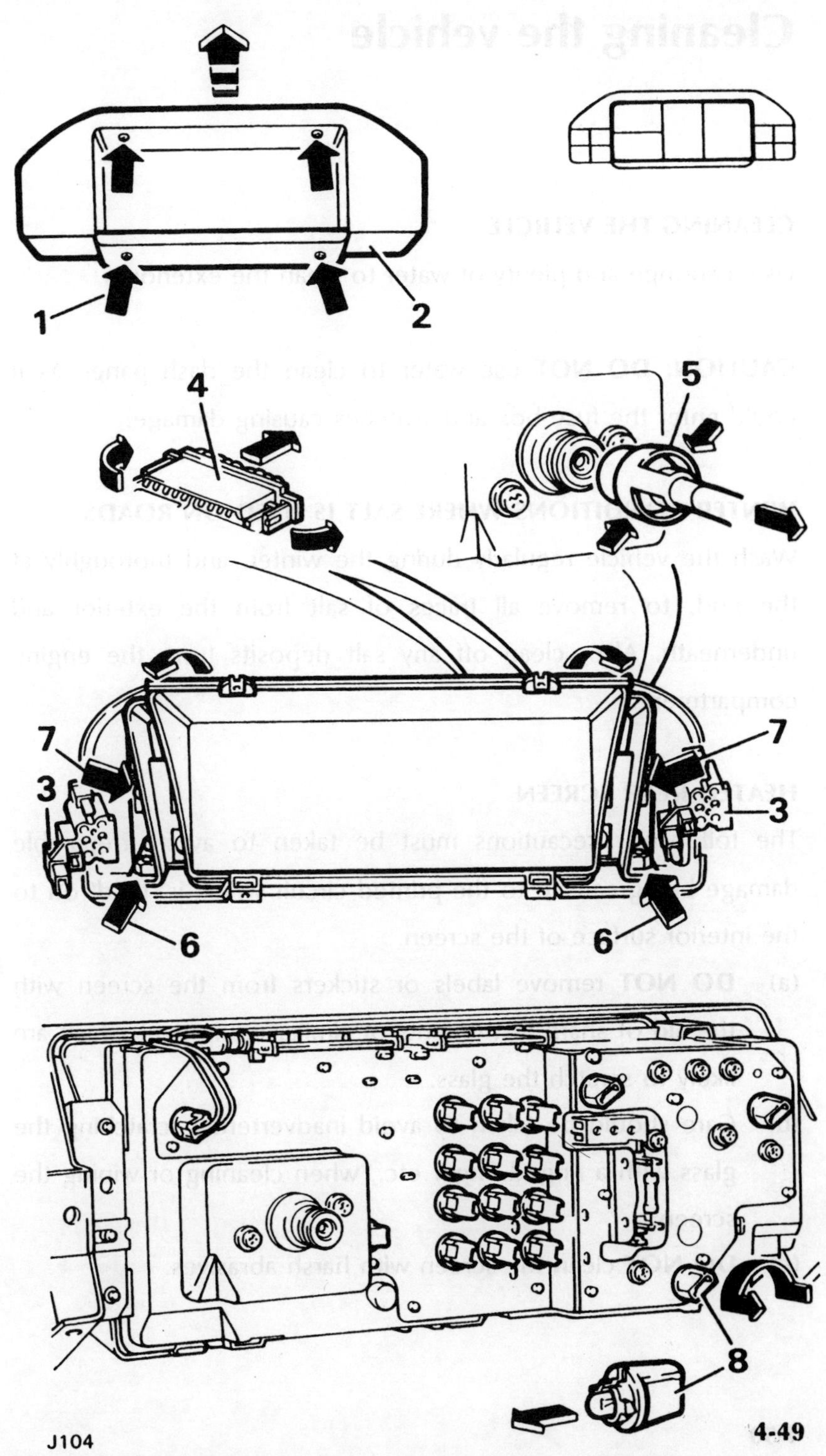

1
2
4
5
7
3
6
7
3
6
8

Cleaning the vehicle

CLEANING THE VEHICLE

Use a sponge and plenty of water to clean the exterior.

CAUTION: DO NOT use water to clean the dash panel, as it could enter the fuse box and switches causing damage.

WINTER CONDITIONS WHERE SALT IS USED ON ROADS

Wash the vehicle regularly during the winter, and thoroughly at the end, to remove all traces of salt from the exterior and underneath. Also, clean off any salt deposits from the engine compartment.

HEATED REAR SCREEN

The following precautions must be taken to avoid irreparable damage being caused to the printed circuit which is 'fired' on to the interior surface of the screen.

(a)   **DO NOT** remove labels or stickers from the screen with the aid of sharp instruments or similar equipment which are likely to scratch the glass.

(b)   Care should be taken to avoid inadvertently scratching the glass with a ringed finger etc., when cleaning or wiping the screen.

(c)   **DO NOT** clean the screen with harsh abrasives.

STEAM CLEANING

To prevent consequential rusting, any steam cleaning within the engine bay must be followed by careful re-waxing of the metallic components affected. Particular attention should be given to the steering column, engine water pipes, hose clips and the ignition coil clamp.

LUBRICATION CHART AND GENERAL DATA

5

Lubricant and fluid capacities

CAPACITIES

The following capacity figures are approximate and are provided as a guide only. All oil levels must be set using the dipstick or level plugs as applicable. Refer to Section 4 for the correct procedure for checking the engine sump.

Engine sump oil,
- 200Tdi models .... 6,00 litres (10.56 pints)
- V8 petrol models .... 5,10 litres (9.00 pints)

Extra when refilling after fitting new filter
- 200Tdi models .... 0,85 litres (1.50 pints)
- V8 petrol models .... 0,56 litres (1.00 pint)

Main gearbox oil (LT77) .... 2,67 litres (4.70 pints)
Transfer box oil .... 2,80 litres (4.90 pints)
Front differential .... 1,70 litres (3.00 pints)
Rear differential .... 1,70 litres (3.00 pints)
Power steering box and reservoir .... 2,90 litres (5.00 pints)
Swivel pin housing oil (each) .... 0,35 litres (0.60 pint)
Fuel tank .... 88,60 litres (19.50 gallons)

Cooling system,
- 200Tdi models .... 11,50 litres (20.20 pints)
- V8 petrol models .... 11,30 litres (20.00 pints)

Diesel engine oil

ENGINE OIL - 200Tdi ENGINES

The minimum performance level oil required for satisfactory engine performance and protection is defined by specifications BLS 22.OL.09 and CCMC PD1.

Oils to BLS 22.OL.09/CCMC PD1

Agip Superdiesel or Sint Turbo Diesel

BP Vanellus C3 or Visco Diesel

Caltex RPM Delo 400*

Castrol Syntron X, TXT, Dynamax or GTX

Century Superb

Duckhams QXR or Hypergrade

Esso Superlube EX 2, Superlube +, Ultra Oil or Super Oil

Gulf Super Diesel or Engine Oil T

Mobil Delvac Super, Mobil 1 Rally Formula or Mobil 1 Formula 15W/50

Kuwait Q8 Auto-4 or Q8 Auto-7

Shell Rimula X or Rotella MTX

Texaco Dieseltex

Oil Viscosity - Ambient Temperatures Applications Chart

SPECIFICATION	SAE VISCOSITY	AMBIENT TEMPERATURE °C
Oils must meet BLS.22.OL.09 or CCMC PD1 or CCMC D3	5W/30 5W/40 5W/50	-30° -20° -10° 0° 10° 20° 30° 40° 50°
	10W/30 10W/40 10W/50	
	15W/40 15W/50	
	20W/40 20W/50	
	25W/40 25W/50	

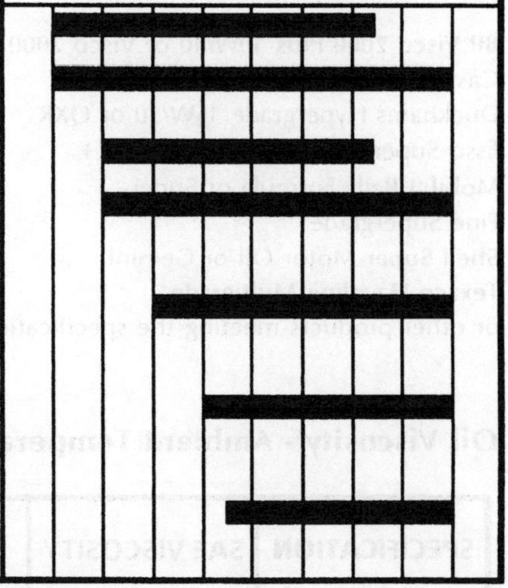

In markets where oils to the above specifications are not available use products to MIL-L-2104D or API CD.

Under severe operating conditions, eg. off road in mud, airborne sand, dust, operating at high speeds in high ambient temperatures above 40°C or continual stop/start operation, the oil and filter change period should not exceed 5000 km (3000 miles). Continuous off road operation in mud, dust and wading conditions requires a monthly oil and filter change. Failure to adhere to the recommended service and operating instructions may result in premature engine wear or damage.

Petrol engine oil

ENGINE OIL - V8 PETROL ENGINES

Recommended lubricants for ambient temperatures above -10°C

BP Visco 2000 Plus 10W/40 or Visco 2000 15W/40
Castrol GTX or TXT or Syntron X
Duckhams Hypergrade 15W/50 or QXR
Esso Superlube Ex2 or Superlube +
Mobil 1 Rally Formula or Super
Fine Supergrade
Shell Super Motor Oil or Gemini
Texaco Havoline Multigrade
or other products meeting the specifications shown in the following chart

Oil Viscosity - Ambient Temperatures Applications Chart

SPECIFICATION	SAE VISCOSITY	AMBIENT TEMPERATURE °C
Oils must meet BLS.22.OL.07 or CCMC G3	5W/30 5W/40 5W/50	
	10W/30 10W/40 10W/50	
Oils must meet BLS.22.OL.02 or CCMC G1 or G2	15W/40 15W/50	
	20W/40 20W/50	
	25W/40 25W/50	

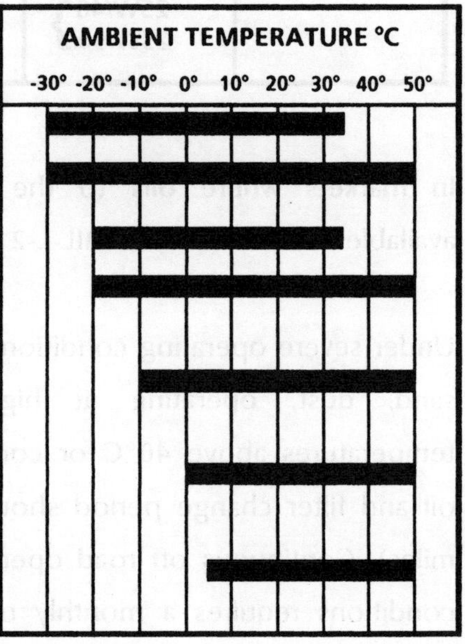

Data - lubricants and fluids

Recommended lubricants and fluids, service instructions for temperate climates - ambient temperature range -10°C to 35°C

COMPONENTS	BP	CASTROL	DUCKHAMS	ESSO	MOBIL	PETROFINA	SHELL	TEXACO
LT77 - five speed gearbox	BP Autran G	Castrol TQF	Duckhams Q-Matic	Esso ATF Type G	Mobil ATF 210	Fina Purfimatic 33G	Shell Donax TF	Texmatic Universal
Transfer box, Final drive units, Swivel pin housing	BP Gear Oil SAE 90EP	Castrol Hypoy 90EP	Duckhams Hypoid 90	Esso Gear Oil GX 85W/90	Mobil Mobilube HD 90	Fina Pontonic MP SAE 80W/90	Shell Spirax 90EP	Texaco Multigear Lubricant SAE 85W/90
Propellor shafts - front and rear, Lubrication nipples (hubs, ball joints etc.)	BP Energrease L2	Castrol LM Grease	Duckhams LB 10	Esso Multi Purpose Grease H	Mobil Grease MP	Fina Marson HTL 2	Shell Retinax A	Marfak All Purpose Grease
Power steering fluid reservoir	BP Autran G	Castrol TQF	Duckhams Q-Matic	Esso ATF Type G	Mobil ATF 210	Fina Purfimatic 33G	Shell Donax TF	Texmatic Universal
Brake and clutch reservoirs	Brake fluids having a minimum boiling point of 260°C (500°F) and complying with FMVSS 116 DOT 4							
Cooling system anti-freeze	Universal anti-freeze, see later page for instructions							

Data - lubricants and fluids

Recommended lubricants and fluids, service instructions all markets

COMPONENTS	BP	CASTROL	DUCKHAMS	ESSO	MOBIL	PETROFINA	SHELL	TEXACO
Seat slides and door lock striker (NLGI-2 multi purpose lithium based grease)	BP Energrease L2	Castrol LM Grease	Duckhams LB 10	Esso Multi Purpose Grease H	Mobil Grease MP	Fina Marson HTL 2	Shell Retinax A	Marfak All Purpose Grease
Windscreen washers	All seasons screen washer fluid							
Bonnet catch	Graphite lock grease type 'B'							
Door locks (anti-burst), Inertia reels	**DO NOT LUBRICATE** These components are 'life' lubricated at the manufacturing stage							
Battery lugs and earthing surfaces where paint has been removed	Petroleum jelly. **DO NOT** use silicone grease							
Air conditioning system refrigerant	**METHYLCHLORIDE REFRIGERANTS MUST NOT BE USED** Use only refrigerant 12. This includes 'Freon 12' and 'Arcton 12'							
Air conditioning system compressor oil	Use Shell Clavus 68, BP Energol LPT 68, Sunisco 4GS, Texaco Capella E Wax Free 68 or Castrol Icematic 99							

Recommended lubricants and fluids, service instructions for ambient conditions outside temperate climate limits or for markets where products listed are not available

COMPONENTS	CLASSIFICATION WORLDWIDE		AMBIENT TEMPERATURE °C
	PERFORMANCE	SAE VISCOSITY	-30° -20° -10° 0° 10° 20° 30° 40° 50°
Front and rear axle differential, swivel pin housings, LT230 transfer box	API GL4 or MIL-L-2105	90 EP or 80W EP	90 EP: bar from -10° to 50°; 80W EP: bar from -30° to 0°
Power steering reservoir, LT77 gearbox	ATF type G		bar from -30° to 50°
Brake and clutch reservoirs	Brake fluid must have a minimum boiling point of 260°C (500°F) and comply with FMVSS 116 DOT 4		bar from -30° to 50°
Lubrication nipples (hubs, ball joints etc.)	NLGI-2 multipurpose lithium based grease		bar from -30° to 50°

Data - anti-freeze

Service instructions for ambient conditions outside temperate climate limits or for markets where the products listed are not available

Anti-freeze

Ethylene Glycol based anti-freeze (containing no methanol) with non-phosphate corrosion inhibitors suitable for use in 200Tdi engines and V8 petrol engines to ensure protection of the cooling system against frost and corrosion.

All engines

One part anti-freeze, one part water, i.e. 50% anti-freeze in coolant. Complete protection below -36°C.

When anti-freeze is not required the cooling system must be flushed out with clean water and refilled with a solution of one part Marstons SQ 36 inhibitor to nine parts water, i.e. 10% inhibitor in coolant.

Engine - 200Tdi models

Bore	90,47 mm (3.562 in)
Stroke	97,0 mm (3.819 in)
Number of cylinders	4
Compression ratio	19.5:1
Cylinder capacity	2495 cc (152 cu in)
Firing order	1, 3, 4, 2
Injection timing	1,54 mm lift at T.D.C.
Tappet clearance, inlet	0,20 mm (0.008 in) - Engine hot
Tappet clearance, exhaust	0,20 mm (0.008 in) - or cold
Valve timing (No. 1 exhaust valve peak)	106° to 109°

Main gearbox - 200Tdi models

Type - Manual	LT77 5-speed helical constant mesh, with synchromesh on all forward gears	
Main gearbox ratios	Fifth (Cruising gear)	0.770:1
	Fourth	1.000:1
	Third	1.397:1
	Second	2.132:1
	First	3.692:1
	Reverse	3.429:1

Data - petrol engines

Engine - V8 petrol models

Bore	88,9 mm (3.500 in)
Stroke	71,12 mm (2.800 in)
Number of cylinders	8
Cylinder capacity	3528 cc (215 cu in)
Compression ratio	8.13:1
Firing order	1, 8, 4, 3, 6, 5, 7, 2
Sparking plug type	Champion N9YC
Sparking plug gap	0,88 to 0,72 mm (0.035 to 0.028 ins)
Distributor	Electronic
Ignition timing, dynamic;	6° BTDC at 750 rpm maximum with vacuum pipe connected using (2 star in UK) 90 minimum octane fuel
Carburetters	Twin S.U. type H.I.F. 44

Main gearbox - V8 petrol models

Type - Manual	LT77 5-speed helical constant mesh, with synchromesh on all forward gears

Main gearbox ratios		
	Fifth	0.770:1
	Fourth	1.000:1
	Third	1.397:1
	Second	2.132:1
	First	3.321:1
	Reverse	3.429:1

Transfer gearbox - all models

Type — LT230T 2-speed reduction on main gearbox output. Front and rear drive permanently engaged via a lockable differential.

Ratios .... High   1.222:1
Low   3.320:1

Rear axle - all models

Type .... Spiral bevel
Ratio 3.538:1

Front axle

Differential .... Spiral bevel
Front wheel drive .... Enclosed constant velocity joint
Ratio 3.538:1
Overall ratio (including final drive)
- all 200Tdi models ....

	high	low
Fifth	3.331:1	9.049:1
Fourth	4.324:1	11.747:1
Third	6.040:1	16.406:1
Second	9.218:1	25.040:1
First	15.962:1	43.367:1
Reverse	14.827:1	40.276:1

Overall ratio (including final drive)
- all V8 petrol models ....

	high	low
Fifth	3.331:1	9.049:1
Fourth	4.324:1	11.747:1
Third	6.040:1	16.406:1
Second	9.218:1	25.040:1
First	14.363:1	39.017:1
Reverse	14.827:1	40.276:1

Data - steering

Steering (lock to lock)

Power assisted .... 3.375 turns
Camber angle .... Zero
Castor angle .... 3°
Swivel pin inclination .... 7°
Front wheel toe-out .... 1,2 to 2,44 mm (0.046 to 0.093 in)

Turning circle between kerbs - all models

205 x 16 tyres .... 11,9 m (39 feet)

Electrical system

Type	Negative earth
Voltage	12
Battery	
- 200Tdi models	643
- V8 petrol models	091
Charging circuit	Alternator
Ignition system	
- V8 cylinder petrol models	Coil

Replacement bulbs and units

Headlamps

- UK and Europe (except France)	60/55 W Halogen bulb
- France	60/55 W Halogen bulb, yellow

NOTE: Local legislative requirements may require fitment of quartz-halogen headlamps in countries outside Europe. Refer to Distributor or Dealer for details.

Front side lamps	12 V 5 W
Side repeater lamps	12 V 5 W
Stop/tail lamps	12 V 21/5 W
Flasher lamps	12 V 21 W
Number plate lamp	12 V 5 W
Reverse lamp	12 V 21 W
Rear fog guard lamp bulb	12 V 21 W
Interior lamp	12 V 5 W
Warning lights (except ignition)	12 V 1.12 W
Ignition warning light	12 V 2 W
Instrument illumination front lighting panel	12 V 1.4 W
Hazard switch warning light	12 V 1.2 W

Data - dimensions and weights

VEHICLE DIMENSIONS - 200Tdi MODELS

Dimensions

Overall length (including spare wheel)	4521 mm (177.9 in)
Overall length (including tow hitch)	4534 mm (178.5 in)
Overall width	1793 mm (70.6 in)
Overall height	1918 mm (75.5 in)
Wheelbase	2540 mm (100 in)
Track front/rear	1486 mm (58.5 in)
Width between wheel boxes	1080 mm (42.5 in)
Seating capacity	5 or 7

Performance

Tyre size fitted	205 R16 radial
Max. gradient (EEC kerb weight)	45°
Approach angle (EEC kerb weight)	42.45°
Departure angle with tow hitch - (EEC kerb weight)	19.76°
Departure angle without tow hitch - (EEC kerb weight)	30.76°
included break over angle	30.79°
Min. ground clearance (unladen)	253 mm (9.9 in)
Wading depth	500 mm (20 in)

Towing weights (refer to Section 3, Towing off-road)

Towing weights	On road	Off road
Unbraked trailers	750 kg	500 kg
Trailers with overrun brakes	3500 kg	1000 kg
4 wheel trailers with coupled brakes - (FULLY BRAKED)*	4000 kg	1000 kg

NOTE: * Only applies to vehicles modified to accept coupled brakes.

IMPORTANT: See NOTE in Section 3, "Towing" for towing a trailer with a weight in excess of 3500 kg.

NOTE: All weight figures are subject to local restrictions.

Data - dimensions and weights

VEHICLE DIMENSIONS - V8 PETROL MODELS

Dimensions

Overall length (including spare wheel)	4521 mm (177.9 in)
Overall length (including tow hitch)	4534 mm (178.5 in)
Overall width	1793 mm (70.6 in)
Overall height	1928 mm (75.9 in)
Wheelbase	2540 mm (100 in)
Track front/rear	1486 mm (58.5 in)
Width between wheel boxes	1080 mm (42.5 in)
Seating capacity	5 or 7

Performance

Tyre size fitted	205 R16 radial
Max. gradient (EEC kerb weight)	45°
Approach angle (EEC kerb weight)	40.63°
Departure angle with tow hitch - (EEC kerb weight)	20.75°
Departure angle without tow hitch - (EEC kerb weight)	31.65°
included break over angle	29.01°
Min. ground clearance (unladen)	241 mm (9.5 in)
Wading depth	500 mm (20 in)

Towing weights (refer to Section 3, Towing off-road)

Towing weights	On road	Off road
Unbraked trailers	750 kg	500 kg
Trailers with overrun brakes	3500 kg	1000 kg
4 wheel trailers with coupled brakes - (FULLY BRAKED)*	4000 kg	1000 kg

NOTE: * Only applies to vehicles modified to accept coupled brakes.

IMPORTANT: See NOTE in Section 3, "Towing" for towing a trailer with a weight in excess of 3500 kg.

NOTE: All weight figures are subject to local restrictions.

Data - vehicle weights

VEHICLE WEIGHTS - 200Tdi MODELS

When loading a vehicle to its maximum (Gross Vehicle Weight), consideration must be taken of the unladen vehicle weight and the distribution of the payload to ensure that axle loadings do not exceed the permitted maximum values.

It is the customer's responsibility to limit the vehicle's payload in an appropriate manner such that neither maximum axle loads nor Gross Vehicle Weight are exceeded.

Maximum EEC kerb weight and distribution - including all optional equipment

Front axle	1037 kg
Rear axle	1043 kg
Total	2080 kg

Maximum axle weights

Front axle	1200 kg
Rear axle	1650 kg
Gross vehicle weight	2720 kg

EEC kerb weight = Unladen weight + Full fuel tank + 75 kg driver.

NOTE: Axle weights are non additive. The individual maximum axle weights nor gross vehicle weight must not be exceeded.

Data - vehicle weights

VEHICLE WEIGHTS - V8 PETROL MODELS

When loading a vehicle to its maximum (Gross Vehicle Weight), consideration must be taken of the unladen vehicle weight and the distribution of the payload to ensure that axle loadings do not exceed the permitted maximum values.

It is the customer's responsibility to limit the vehicle's payload in an appropriate manner such that neither maximum axle loads nor Gross Vehicle Weight are exceeded.

Maximum EEC kerb weight and distribution - including all optional equipment

Front axle .... 961 kg
Rear axle .... 1018 kg
Total .... 1979 kg

Maximum axle weights

Front axle .... 1100 kg
Rear axle .... 1650 kg
Gross vehicle weight .... 2720 kg

EEC kerb weight = Unladen weight + Full fuel tank + 75 kg driver.

NOTE: Axle weights are non additive. The individual maximum axle weights nor gross vehicle weight must not be exceeded.

Data - fuel economy

FUEL ECONOMY (Passenger Car Fuel Consumption Order

1983 No. 1486 80/1268 EEC)

200Tdi models

Simulated urban cycle (mpg)	30.5
Constant speed, 56 mph (mpg)	42.4
Constant speed, 75 mph (mpg)	28.9
Simulated urban cycle (l/100 km)	9.3
Constant speed, 90 kph (l/100 km)	6.7
Constant speed, 120 kph (l/100 km)	9.8

V8 petrol models

Simulated urban cycle (mpg)	13.0
Constant speed, 56 mph (mpg)	26.2
Constant speed, 75 mph (mpg)	19.6
Simulated urban cycle (l/100 km)	21.7
Constant speed, 90 kph (l/100 km)	10.8
Constant speed, 120 kph (l/100 km)	14.4

The above results were achieved under controlled test conditions in compliance with the Order, and do not express or imply any guarantee of the fuel consumption of any particular vehicle with which this information may be supplied. Vehicles are not individually tested, and there are inevitably differences between individual vehicles of the same model. In addition, the vehicle may incorporate particular modifications. Furthermore, the driver's style and road traffic conditions, as well as the extent to which the vehicle has been driven and the standard of maintenance will all affect its fuel consumption. Information as to the results of officially approved tests on all vehicles tested is available for inspection by customers on the premises where these vehicles are displayed.

Data - tyre pressures

⚠️ **WARNING: Tyre pressures must be checked with the tyres cold, as the pressure is about 0.21 bar (3 lbf/in²) 0.2 kg/cm² higher at running temperature. If the vehicle has been parked in the sun or high ambient temperatures, DO NOT reduce the tyre pressures, move the vehicle into the shade and wait for the tyres to cool before checking the pressures.**

TYRE PRESSURES

Maximum tyre life and performance will only be obtained if the tyres are maintained at the correct pressures.

	Front	Rear
Normal - all load conditions	1,9 bar	2,6 bar
	28 lbf/in²	38 lbf/in²
	2,0 kgf/cm²	2,8 kgf/cm²
Emergency soft	1,2 bar	1,7 bar
	17 lbf/in²	25 lbf/in²
	1,2 kgf/cm²	1,7 kgf/cm²

NOTE: Emergency soft pressures should only be used in extreme conditions where extra floatation is required and maximum speed must be restricted to 40 km/h (25 mph). Return pressure to normal immediately firm ground is regained. When the vehicle is used for towing, the reduced tyre pressures are not applicable.

PARTS & ACCESSORIES 6

Nudge Bars

An integrated vehicle and accessory design philosophy has resulted in this attractively styled family of nudge bars. Testing in the heat of the Australian desert, the cold of the Canadian tundra, and in the laboratory has ensured that their performance is compatible with the considerable capabilities of your Discovery. Manufactured to exacting Land Rover standards to ensure strength and durability, the bars are encapsulated in a hardwearing dipped nylon coating, and incorporate positions for mounting auxiliary lights.

RTC9500 - Nudge bar, with removeable lamp guards.

RTC9501 - Nudge bar.

RTC9502 - Nudge bar, 'A' line.

Hinged Lamp Guards

These uniquely designed front and rear lamp guards protect lenses from flying stones and minor impacts. Horizontal slats, which are so positioned to allow maximum light transmission, are manufactured from black nylon coated steel to give strength and longevity. The guards can be removed quickly to make short work of lamp cleaning.

RTC9504 - Set of four.

RTC9503 - Rear pair.

Parts & accessories

Auxiliary Lighting

Penetrating quartz halogen driving and fog lamps will see you through any conditions on or off road. The round Rally lamps have been proven in a wide range of events around the world. The rectangular lamps use computer age projector technology to give a precise, powerful beam from a small package.

RTC8914 - Rally 2000 driving lamp.

RTC9521 - Rally 1000 fog lamp.

RTC9522 - Rally 1000 driving lamp.

RTC9523 - Projector fog lamp pair.

RTC9524 - Projector driving lamp pair.

Lighting Pods

A pair of high technology fog lamps, incorporated into purpose designed nacelles give improved lighting in conditions of poor visibility. The smooth, integrated appearance of this installation on the front spoiler gives the best position for maximum effectiveness. The beams can be aligned easily from the front of the units.

RTC9525 - Projector lighting pods, fog pattern, pair.

Side Moulding Protector Kit

Resilient rubber compound protective finishers, combined with ABS wheelarch covers offer bodywork protection from car park scrapes and minor impacts. Discovery side finishers are self-adhesive, and require no drilling of the bodywork.

RTC9532 - Side moulding protector kit, vehicles without stripe treatments.

Stowable Loadspace Protector

Tough liners of wipe clean, PVC coated nylon protect the loadspace area - including the rear door trim. The design still allows the use of side facing rear seats when this option is fitted. Retention by Velcro allows easy removal; the protector can be folded to a compact size when not needed.

RTC9531 - Stowable loadspace protector.

Rigid Loadspace Protector

A liner made from shatterproof ABS will take just about anything which is thrown at it. Moulded to fit the counters of the loadspace, an anti-slip mat is included with the liner. The practical slate colour is matched to the interior of your Discovery.

RTC9530 - Rigid loadspace protector.

Parts & accessories

Roller Loadspace Cover

A nylon cover screens your luggage from prying eyes and screens it from the sun. The cover retracts on to a compact spring loaded roller in seconds, and can be removed simply.

RTC9561 - Roller loadspace cover.

Dog Guards and Gun Clips

A snug fitting dog guard effectively separates the load area from the passenger compartment, but can be removed easily when necessary. The bar pattern guard looks at home within the Discovery interior, being colour keyed in a special gunmetal nylon finish. A mesh style guard, still utilising the contour hugging frame design, is also available. Gun clips can attached to both types of dog guard. They enable up to 4 guns to be carried in their foam padded supports.

RTC9508 - Grey bar.

RTC9509 - Grey mesh.

RTC9510 - Gun clips for bar dog guard.

RTC8108 - Gun clips for mesh dog guard.

Luxury Carpet Mats

A carpet underfoot which combines bespoke, contour perfect, tailoring with a deep Wilton pile sounds an impossible luxury for an off-road vehicle. Weave the carpet in a tough but washable ANSO IV nylon, incorporate a hardwearing heelpad, provide an anti-slip backing, and the seemingly conflicting objectives of comfort and protection are beautifully achieved.

RTC9511 - Footwell set, RHD.

RTC9512 - Footwell set, LHD.

RTC9513 - Rear loadspace.

Rubber Mats

Thoughtful design has resulted in these moulded rubber mats; a perimeter wall prevents any spillage of mud or water on to the carpet. The Discovery motif is incorporated into the unique stud pattern, which features a cleverly concealed heel pad for the heavy wear areas. The practical slate finish tones with the interior trim colours.

RTC9514 - Footwell set.

Parts & accessories

Waterproofs

Covers which are truly waterproof are essential items to protect front and rear seats. After use they can be wiped clean and then stored in their carry bag - ready for the next occasion when your Discovery interior needs protection from the elements.

RTC9516 - Front seat, pair.

RTC9517 - Front and rear seat, set.

RTC9518 - Inward facing seat, pair.

Alloy Wheels

Specify a set of these five spoke alloy wheels to personalise the appearance of your Discovery. The 7J profile accepts the range of approved tyres, and has the rugged strength to survive tough off road conditions.

RTC9526 - Styled alloy road wheel.

Stripe Treatments

These accessories have been designed by the same Land Rover stylists who crafted the shape of Discovery, which is why the bodywork and stripes appear as natural partners. The strobe additions gives even greater individuality to the vehicle.

RTC9554 - Stripe treatment, blue keyline.

RTC9555 - Stripe treatment, green keyline.

RTC9559 - Strobe stripes, for RTC9554.

RTC9560 - Strobe stripes, for RTC9555.

Spare Wheel Cover and Case

A spare wheel cover with the Discovery signature provides the finishing touch to the appearance of your vehicle. The rigid case is made from a tough ABS material - and three catches allow access to the spare wheel in seconds. The elasticated vinyl cover features matching white piping and graphics.

RTC9528 - Rigid spare wheel case.

RTC9529 - Vinyl spare wheel cover.

Mudflaps

Heavy duty, moulded rubber front mudflaps reduce spray and protect vehicle paintwork.

RTC9562 - Front mudflap kit.

Locking Wheel Nuts

A set of locking wheel nuts can be a good investment when compared with the cost and inconvenience of the theft of wheels and tyres. These wheel nuts adhere to the safety critical standards which are applied to the standard items on the vehicle. Unlike an ordinary lock, the pin drive 'key' system has no delicate parts to give problems in service.

RTC9535 - Steel wheel set of five.

RTC9533 - Alloy wheel set of five.

Parts & accessories

Baby Seat

The Land Rover baby seat is suitable for use from birth to approximately 4 years of age. Used rearward facing, babies up to 20 lb (9 kg) in weight can be held securely on the front passenger seat. Turn the seat round for children between 20 and 40 lb (9 kg to 18 kg) to be accommodated on either front or rear seats. The seat is restrained using the Discovery lap and diagonal seat belts, and a pair of special straps are supplied for the centre rear seat position. The smart pinstripe cover is removable, and machine washable. Approved to BS AU 202, BS 3254 and EEC standards.

RTC9534 - Baby seat.

Fire Extinguishers

A fire extinguisher ensures peace of mind when driving under diverse and extreme environments. These BCF type extinguishers use a Halon gas which is effective against the types of fire likely to be found in a vehicle, and it does not leave a residue after use. Strong die cast canisters are fitted with safety interlocks to prevent accidental discharge. The 2,0 kg extinguisher features a contents gauge, and a quick release 'safety belt' fastening.

RTC9537 - 1,0 kg.

RTC9536 - 2,0 kg.

Warning Triangle

A mandatory continental requirement, the folding triangle is light in weight, sturdy, and highly reflective by day and night. Folds away into a space saving moulded storage box.

RTC8937 - Warning triangle.

Equipment Stowage

You can provide extra stowage for valuables and the other smaller items which accumulate within a vehicle. The cubby box and the soft luggage bag are designed to fit on the centre console - within easy reach of the driver and front seat passenger. The lockable panniers locate within the loadspace side pockets to convert these into secure areas.

RTC9552 - Lockable cubby box, centre console.

RTC9553 - Lockable panniers.

RTC9556 - Soft luggage bag.

Parts & accessories

Cockpit Handlamp

Ideal for map reading, the halogen bulb of the cockpit handlamp is powered from the cigar lighter socket. A dashboard mounting bracket is included, and permanent magnets built in to the back of the lamp body increase its versatility. The cable will extend from 400 mm to 3000 mm to allow the lamp to be used as a compact worklight.

RTC9527 - Cockpit handlamp.

Sunhatch

Let the sun shine through this original equipment pattern sunhatch. The screened glass hatch can be tilted for extra ventilation, or be removed completely. The kit is supplied complete with a wind deflector and stowage bag.

RTC9558 - Sunhatch kit, front or rear.

Touch-Up Paint

Stone chips and minor abrasions can be covered by use of colour matched touch-up paint. Popular colours are available in both pencil and aerosol versions.

	Aerosol	Pencil
Davos	RTC4058A	RTC4058T
Windjammer	RTC6401A	RTC6401T
Corallin	RTC5728A	RTC5728T
Caracal	RTC5979A	RTC5979T
Zanzibar	RTC5729A	RTC5729T
Marseilles	RTC6400A	RTC6400T
Foxfire	RTC5983A	RTC5983T
Mistral	RTC6402A	RTC6402T
Arken	RTC6454A	RTC6454T

Inward facing seat

Increase the passenger carrying capability of your Discovery by fitting one or two inward facing seats. Supplied complete with lap safety belt, the seat is trimmed in co-ordinated Sonar blue, and retracts neatly when not in use.

RTC9557 - Inward facing seat.

Parts & accessories

Roof Rack System

This roof rack system extends the carrying potential of your Discovery. The provision of a full load bed, and roof bars within the roof rack frame allows the transportation of sports equipment or a wide variety of differing loads up to a weight of 75 kg (165 lbs). The rear mounted ladder adds style as well as accessibility.

RTC9539 - Roof rack.

RTC9540 - Ladder.

RTC9477 - Load stops.

RTC9478 - Quick strap.

Angled Ski Carrier

A new design of carrier will hold up to six pairs of skis, angled to make sure that bindings will not catch the roof or each other. The lockable arm is spring loaded to allow ease of use.

RTC9544 - Angled ski carrier, for use with roof rack, or vehicle option roof bars.

Sailboard Carrier

Two specially moulded rubber crossbar covers together with a pair of sophisticated terylene straps ensure that sailboards are supported securely.

RTC9542 - Sailboard carrier, for use with roof rack, or vehicle option roof bars.

Rear Ski Carrier

A purpose designed carrier is fastened securely to the the spare wheel, allowing easy access to up to four pairs of skis. The locks provided ensure that security is maintained.

Rear Step

The retractable step gives an easy entrance and exit at the rear of your vehicle. A novel, patented design concept utilises a gas strut to achieve controlled retraction in an effortlessly smooth fashion. This new product will appeal to all members of the family.

RTC9505 - Automatic rear step.

Side Runners

Making an entrance into your Discovery will be a stylish event when you specify a pair of side runners - and they will add yet more style to the appearance of the vehicle. The tread insert utilises the unique stud pattern to give a sure footing whatever the conditions.

RTC9507 - Side runners, pair.

Parts & accessories

Alarm System

The installation of an alarm system has become an almost mandatory safeguard against theft and vandalism. This system has been developed specifically for Land Rover; advanced microprocessor and ultrasonic technology was chosen carefully to give maximum protection levels. A comprehensive test programme, whilst checking the reliability and effectiveness of the alarm, also ensured that the electrical integrity of the vehicle has been maintained. As a further benefit the alarm remote control can be made to operate central locking, and close electric windows, from up to 30 metres away - a real bonus for convenience.

RTC8895 - Remote control alarm system.

RTC9546 - Remote control alarm system, with central locking interface.

RTC8894 - Window closer interface, for use with RTC9546.

- Twin siren - 124 dB at one metre.

- All indicators flash when the alarm sounds.

- Ultrasonics detect motion within the vehicle.

- Current drain circuit triggers the alarm if a door, the bonnet or the tailgate is opened.

- Ignition immobilisation circuit is effective with both petrol and diesel engines.

- Safety interlock circuit stops the system being switched on whilst your Discovery is running.

Cellular Telephones

The Land Rover cellular telephone allows you to keep your finger on the of the pulse of business; there is even a transportable version which can be used when you are out of your Discovery. The telephone incorporates a hands free facility, a 40 number memory, and ETACs, to make calls as easy as possible. A combination of full Land Rover approval, a two year equipment warranty, with a reputable airtime supplier gives the assurance of an effective cellular installation.

RTC9548 - Cellular telephone.

RTC9547 - Cellular telephone with transportable facility kits to allow the transfer of cellular telephones, and the installation in a second vehicle are also available via your local dealer.

Towing Equipment

Tow Bars

A superlative towing capability can be harnessed with the fitment of the multi-height bar. Useful when more than one type of trailer is likely to be towed, the kit is supplied complete with tow ball, electrics, protective covers - even a tube of grease and gloves.

RTC9499 - Multi-height tow bar kit, including tow bar, 50 mm tow ball, 'N' type towing electrics, tow ball and socket covers, tube of grease and pair of gloves.

Parts & accessories

Combination Tow Ball/Jaw

The combination ball and jaw unit is the best answer where trailers with different hitches need to be towed. Both the 50 mm ball and the pin are rated to pull up to 3,5 tonnes.
RTC8159 - Combination tow ball/jaw unit, 3,5 tonne capacity.

Towing Electrics

Discovery towing electrics kits are BSI approved, and are complete with 7 pin socket, cables and fixings. The addition of an S type auxiliary socket kit allows power to be fed into a caravan or trailer for equipment such as a refrigerator or interior lights. The split charge unit permits the electrical system within your Discovery to charge both the vehicle and auxiliary batteries.
RTC8978 - S type auxiliary socket kit.
RTC8977 - Split charge unit.

Tow rope

A strong rope for towing, coupling to vehicles being facilitated by a sturdy hook bound on to each end. Manufactured from buoyant, "safety" red polypropylene, the rope is cleaned easily after use.

Parts & accessories

Husky Electric Winch

A top quality electric winch for self-recovery or general winching, the Husky has been designed for neat installation under the front bumper. Ideal for the retrieval of horseboxes and trailers, or the hauling of boats on to slipways, the unit also has the capacity for professional winching requirements.

- Fully reversible power for two way winching.

- Remote control with 3,7 metre (12 ft) wanderlead.

- Fully automatic load actuated braking system.

- Sealed for life, water resistant winch gear housing.

- Rated capacity 3850 kg (8500 lb), installed capacity 2040 kg (4500 lb).

RTC9519 - Husky electric winch, with fitting kit, rope, hook, remote control and all fittings.

Winch Accessory Kit

A canvas stowage bag contains the essential winching accessories of snatch block, strap, gloves and a guide to winching techniques.

RTC9520 - Winch accessory kit.

Please consult your local Land Rover dealer for availability and territorial applicability.

OFF-ROAD DRIVING

7

DRIVING OFF-ROAD

This section is to help you take advantage of the outstanding ride comfort of the Discovery and to realise the off-road potential of the vehicle. Obviously more stress is imposed upon a vehicle that is being driven off-road, but the effect of this can be minimised by using the correct driving technique.

This is a guide to the more important aspects of off-road driving. The techniques described on the following pages will with practice enable the driver to get maximum benefit from the Discovery. It is important that the driver is familiar with the vehicle and to ensure that it is properly maintained.

⚠️ **WARNING: Driving off-road can be hazardous. DO NOT take unnecessary risks. Drive carefully and be prepared for emergencies.**

Off-road driving

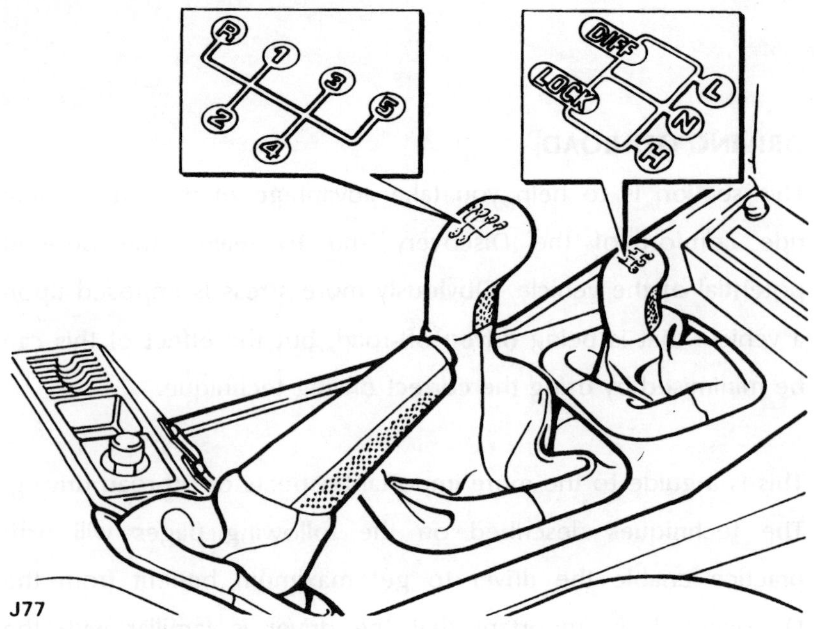

Gear selection changing

For off-road driving, selection of the correct gear range (H or L) and the correct main gear is extremely important. Only experience will tell you which is the correct gear for a given section of ground, but generally the higher the gear the better. **DO NOT** change gear or declutch while negotiating difficult terrain as the drag on the wheels may cause the vehicle to stop when the clutch is depressed and you may experience difficulty in restarting. High range should be used whenever possible, only change to low range when ground conditions become very difficult. The **DIFF LOCK** should be engaged where there is any risk of loosing wheel grip and unlocked as soon as firm, level non-slippery ground is reached.

Slipping the clutch

Use of excessive clutch slip to prevent the engine stalling will result in premature clutch wear. A low enough gear should be selected to enable the vehicle to proceed without slipping the clutch.

DO NOT drive with your foot resting on the clutch pedal as this could result in loss of control of the vehicle by inadvertantly depressing the clutch pedal as the vehicle travels over a sudden bump.

Braking

Keep the application of the brake pedal to an absolute minimum. Braking on wet, muddy or loose surfaces will almost certainly cause one or more wheels to lock and the resulting slide could prove dangerous.

Accelerating

Use the accelerator with care as sudden power surges may induce wheel spin and result in loss of control of the vehicle.

Off-road driving

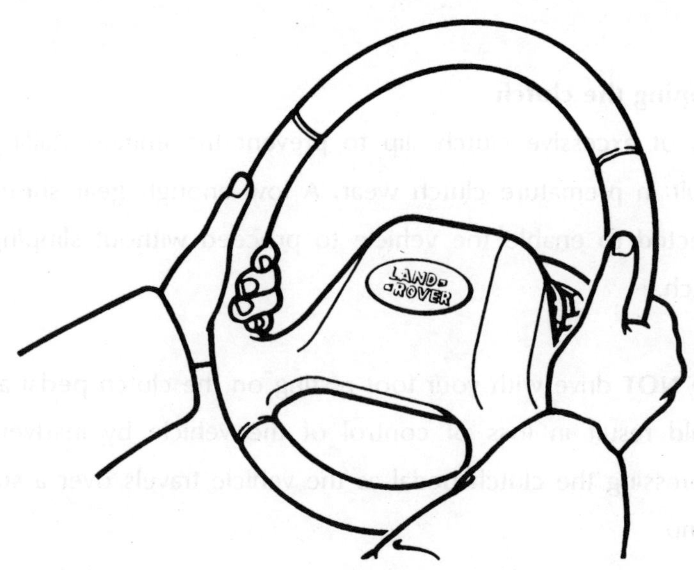

J58

⚠ **WARNING: DO NOT hold the steering wheel with the thumbs inside the rim. A sudden violent kick of the wheel could damage or even break the thumbs. Grip the wheel on the outside of the rim when travelling across country. Fig. J58**

Survey on foot before driving.

Before negotiating a difficult section of terrain, you should carry out a preliminary survey of the ground on foot in order to minimise the risk of getting caught in a previously unnoticed hazard.

J59

Driving on soft ground - Fig. J59

When driving on soft ground as high a gear as practical should be selected. If there is any risk of loosing wheel grip the **DIFF LOCK** should also be engaged. For extremely difficult conditions, reduced tyre pressures will increase the contact area of the tyres with the ground and improve wheel grip. Remember that reduced tyre pressures will also reduce the vehicles ground clearance. Tyre pressures **MUST** be returned to normal as soon as possible. Refer to Data Section for maximum and minimum permitted tyre pressures. Unlock the **DIFF LOCK** as soon as firm non-slippy ground is reached.

Off-road driving

J62

Soft dry sand - Fig. J62

Select a suitable gear and **STAY** in that gear until a firm surface is reached. It is generally advisable when driving in soft sand to use low range as this will enable you to accelerate through suddenly worsening conditions without the risk of being unable to restart, having stopped to change from high to low range. The **DIFF LOCK** should be engaged. Remember in soft conditions that reduced tyre pressures will increase the contact area with the ground but reference should first be made to ascertain the correct tyre pressures for the prevailing conditions. If the tyre pressures have been reduced for soft ground conditions they must be reinflated upon regaining firm ground.

Gear changing should not be attempted as depressing the clutch to change gear in soft sand will cause the vehicle to stop because of the drag at the wheels.

When stopping your vehicle in sand remember that re-starting while facing up a slope is almost impossible and you should therefore park on level ground, or with the vehicle facing downhill. In order to avoid wheel spin a standing start is best achieved using second or third gear, and the minimum throttle opening that is necessary to start moving. If forward motion is lost avoid excessive wheel spin as this can only make things worse. Clear the sand from the tyres and ensure that the chassis and axles are not touching the sand.

If the wheels have sunk deep into the sand it will be necessary to lift the vehicle using an air bag or high lift jack and then build up the sand under wheels so that the vehicle, when lowered, will be on level ground. If a restart is still not possible it may be necessary to place sand mats or ladders beneath the wheels. Great care must be taken to avoid personal injury when jacking the vehicle. Disengage the **DIFF LOCK** as soon as firm ground is reached.

Off-road driving

J61

Ice and Snow - Fig. J61

The driving techniques employed in snow and icy conditions are generally similar to those used for driving on mud or wet grass. Select the highest gear possible with the **DIFF LOCK** engaged and drive away using the minimum of throttle opening. Avoid violent movements of the steering wheel and keep braking to a minimum. Drive slowly and **DO NOT** brake hard. Remember to disengage the **DIFF LOCK** as soon as non icy ground is reached.

Driving on rough tracks - Fig. J60

Although beaten rough tracks can be negotiated in normal drive, it is advisable to lock the differential if there is excessive suspension movement likely to induce wheel spin. As the track becomes rougher it may be necessary to engage low range to enable a steady low speed to be maintained with overrun engine braking without constant use of the brake and clutch pedals. Unlock the **DIFF LOCK** as soon as smooth firm ground is reached.

Off-road driving

J65

Climbing Steep Slopes - Fig. J65

When climbing or descending steep slopes always engage **DIFF LOCK** and follow the fall line, as travelling diagonally may result in the vehicle sliding broadside down the slope. When climbing steep slopes, particulary if the surface is loose or slippery, use sufficient speed in the highest practical gear, because this enables the driver to take advantage of vehicle momentum. Too much speed when climbing a hill with a bumpy surface can result in one or more wheels lifting, causing the vehicle to loose traction and stop. In this case a slower approach may be more successful. Traction can be improved by easing off the accelerator just before loss of forward motion.

If the vehicle fails to climb a hill but does not stall, the following procedure should be adopted:-

(a)   Hold the vehicle on the foot brake. It will be necessary to use the handbrake only if the foot brake fails to hold due to wet brake linings.

(b)   Engage reverse gear low range as quickly as possible.

(c)   Release the clutch and brakes simultaneously.

(d)   Allow the vehicle to reverse down the slope using engine over run braking to control the speed of the decent. Accelerate slightly if the vehicle begins to slide down the hill.

(e)   **DO NOT** apply the brake or clutch pedal during the decent. Even a light application may cause the front wheels to lock and this would render the steering ineffective.

If the engine stalls while climbing the hill, the following procedure is recommended:-

(a)   Hold the vehicle on the foot brake and hand brake.

(b)   Engage reverse gear low range, release the handbrake and remove feet from brake and clutch pedals.

(c)   Start the engine in gear and allow the vehicle to reverse down the hill using engine over run braking to check the speed. A laden vehicle on a steep hill will start without the aid of the starter motor as soon as the brakes are released, if there is sufficient traction.

Off-road driving

When back on level ground or where traction can be regained, a faster approach and the resulting extra momentum will probably enable the hill to be climbed.

If the hill is too difficult to climb, **DO NOT** take unnecessary risks, try to find an alternative route. Only experience will give you the correct approach for each new hill climb.

Ground Clearance

Don't forget to allow for ground clearance under the chassis and axle differentials and under the front and rear bumpers of the vehicle at the bottom of a steep slope. Avoid deep wheel ruts, sudden changes in slope and obstacles which may foul the chassis or axles. On soft ground the axle differentials will clear their own path in all but the most difficult conditions. However, on frozen, rocky or dry, hard ground, hard contact between the differentials and the ground will generally result in the vehicle coming to a sudden stop.

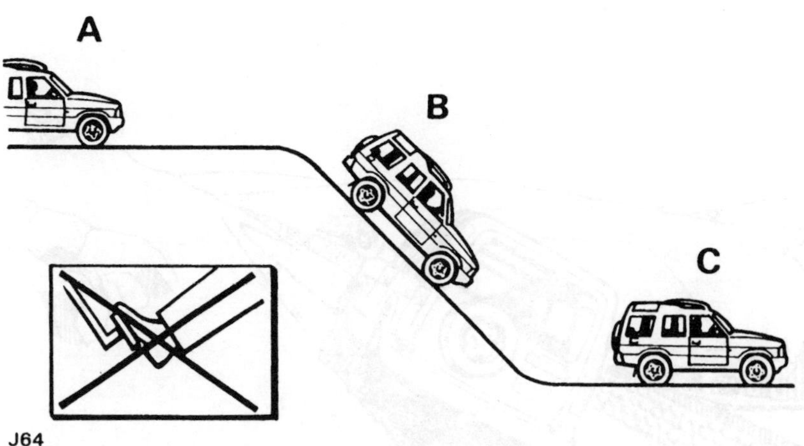

Descending steep slopes - Fig. J64

Stop the vehicle at least a vehicles length before the slope and engage first gear, low range with the differential locked. Check gear engagement before moving off. **DO NOT** touch the brake or clutch during the descent - the engine will limit the speed, and the vehicle will remain perfectly under control while the front wheels are turning. If the vehicle begins to slide, accelerate gently to maintain directional stability.

(a)   Stop at least a vehicles length before the slope. Select first gear low range with the differential locked.

(b)   Engine retardation

(c)   Now unlock differential and change into second or third gear.

Off-road driving

Traversing a slope - Fig. J70

Traversing a slope should be undertaken having observed the following precautions:-

- Check that the terrain is firm under all wheels and that the ground is not slippery.

- Check that the downhill wheels are not likely to drop into a sudden depression in the ground as this will suddenly increase the angle of tilt.

- For the same reason ensure that the uphill wheels do not run over rocks, tree roots, or similar obstacles.

- Any load carried in the back of the vehicle should be evenly distributed as low as possible and made secure. A sudden shift of load while traversing a slope could cause the vehicle to overturn.

⚠️ **WARNING: Any roof rack luggage should be removed, as a high positioned load could cause the vehicle to roll over. Failure to follow these instructions may cause the vehicle to roll over !!**

Off-road driving

J71

Negotiating a 'V' shaped gully - Fig. J71

This should be tackled with extreme caution, as steering up one or other of the gully walls could lead to the vehicle being trapped with its side against the opposite gully wall.

J66

Existing Wheel Tracks - Fig. J66

Avoid over-steering while driving along wheel tracks. Otherwise this could result in the vehicle being driven on full left or right hand lock in the ruts. This must be avoided as it causes drag at the front wheels and is especially dangerous because it can result in the vehicle suddenly veering off the track when the front wheels reach level ground or traction is found. Do not "fight" the vehicle. Let it steer itself along the rut as much as possible, keeping a firm grip of the steering wheel, do not let it spin free.

Off-road driving

J67

Crossing a ridge - Fig. J67

Always approach a ridge at right angles so that both front wheels and then both rear wheels cross together. If approached at an angle traction can be lost completely through diagonally opposite wheels lifting off the ground.

J68

Crossing a Ditch - Fig. J68

Ditches should always be crossed at an angle so that three wheels maintain contact with the ground assisting the passage of the fourth wheel across the ditch. If approached straight on, both front wheels will drop into the ditch probably with the chassis and the front bumper trapped on opposite sides of the ditch. It is advisable to have the **DIFF LOCK** engaged.

Off-road driving

Wading - Fig. J69

The maximum advisable wading depth is 0,5 metre. Before negotiating a deep water crossing it is recommended that a drain plug is fitted to the diesel timing cover and the V8 clutch housing - see Section 4. If the water depth exceeds 0,5 metre a sheet of plastic or other water resistant material draped in front of the radiator grille to prevent any water from passing through, will reduce the risk of saturation of the engine.

NOTE: Drain plugs are not supplied with the vehicle but are available from the Parts Department at your Land Rover Dealer.

Off-road driving

A plastic sheet will also help to prevent mud and floating debris blocking the radiator. Land Rover drivers throughout the world frequently travel through water where the depth exceeds 0.5 metre having taken the following precautions:-

(a)   Generally stagnant water is more likely to be a hazard than a river or stream as flowing water tends to prevent a build-up of silt. The silt in a stagnant pool can be several feet deep. Always ensure that the river or pool bed is firm enough to support the weight of the vehicle and provide traction.

(b)   Ensure that the engine air intake is kept clear of the water.

(c)   A low gear is desirable with the **DIFF LOCK** engaged, and sufficient throttle should be maintained to avoid stalling the engine if the exhaust is under water.

(d)   Drive slowly into water and accelerate to a speed which causes a bow wave to form and maintain that speed.

Off-road driving

After wading

Make sure that the brakes are dried out immediately so that they are fully effective when needed again. This can be accomplished by driving a short distance with the foot brake applied. **DO NOT** rely on the handbrake to hold the vehicle once the transmission brake has been subjected to mud and water; leave the vehicle parked in gear.

On V8 petrol models remove the clutch housing drain plug and on 200Tdi models remove the timing cover drain plug, also any covering material from the front of the radiator grille. If the water was particularly muddy it is possible that the radiator may be blocked with mud and leaves and this should be removed immediately to reduce the risk of overheating. If deep water is regulary negotiated it would be wise to check all transmission oils for signs of water contamination after each trip. Contaminated, emulsified oil can be easily recognised by its milky appearance.

AFTER CROSS COUNTRY DRIVING

If the tyre pressures have been reduced, they **MUST** be restored to the normal recommended pressures as soon as reasonable road conditions or hard ground is reached.

⚠️ **WARNING: Before rejoining the highway (public metalled roads) or driving the vehicle at speeds above 40 km/h (25 mph), wheels and tyres must be inspected for damage, after cleaning off any mud. Do not forget the inside faces. Also, inspect the brake discs and calipers, and remove any stones or grit etc., that may have become lodged and affect the efficiency of the brakes.**

Vehicle recovery

VEHICLE RECOVERY

Should the vehicle become immobile due to loss of wheel grip, the following hints could be of value:

(a) Avoid prolonged wheel spin; this will only make matters worse.

(b) Try to remove obstacles rather than force the vehicle to cross them.

(c) If the ground is very soft, reduce tyre pressures if this has not previously been done.

(d) Clear clogged tyre treads.

(e) Reverse as far as possible, then attempt an increased speed approach. The momentum reached in going forward again may get the vehicle over the obstacle.

Typeset and printed by Land Rover Printing Services (11/89)

LAND ROVER OFFICIAL FACTORY PUBLICATIONS

Land Rover Ser. 1 Workshop Manual   4291
Land Rover Ser. 1 1948/53 Parts Catalogue   4051
Land Rover Ser. 1 1954/58 Parts Catalogue   4107
Land Rover Ser. 1 Instruction Manual   4277
Land Rover Ser. 1 & II diesel Instruction Manual   4343

Land Rover Ser.II & IIA Workshop Manual (part 1 engine)   AKM8159
Land Rover Ser.II & IIA Workshop Manual (part 2 chassis)   AKM8159
Land Rover II/IIA bonneted control Parts Catalogue   605957
Land Rover Ser. IIA Parts Catalogue   RTC9840CC
Land Rover IIA/IIB Instruction Manual   LSM641M

Land Rover Ser. III Workshop Manual   AKM3648
Land Rover Ser. III Workshop Manual V8 Supplement (edn. 2)   AKM8022
Land Rover Ser. III Parts Catalogue   RTC9841CE
Land Rover Ser. III Owners' Manual (1971-81)   607324B
Land Rover Ser. III Owners' Manual (1981-85)   AKM8155

Land Rover 90/110 & Defender Workshop Manual   SLR621ENWM
Land Rover 90/110 Owners' Manual   LSM0054

Land Rover Discovery Workshop Manual   SJR900ENWM
Land Rover Discovery Owner's Handbook (1989 on)   SJR 820 ENHB 90

Land Rover Military (Lightweight) Ser. III Parts Catalogue
Land Rover Military (Lightweight) Ser. III User Manual   608180B
Land Rover 101 1 tonne forward control Workshop Manual (soft & hard cover) RTC9120B
Land Rover 101 1 tonne forward control Parts Catalogue   6082943
Land Rover 101 1 tonne forward control User Manual   608239

Range Rover Workshop Manual (edn. 7) (1970-85)   AKM3630
Range Rover Workshop Manual 1986 on (includes LSM 180WM)   SRR660EN WM
Range Rover (up to 1985) Parts Catalogue   RTC9846CH
Range Rover (2 door) Owners' Handbook   606917
Range Rover Owners Handbook (1981 on)   AKM 8139
Range Rover Owners Handbook (1983 on)   LSM 0001HB

Land Rover and Range Rover Driving Techniques   LR 369
Land Rover's Manual for Africa   SMR 684 MI

Owners' Workshop Manuals
Land Rover 2/2A/3 1959-83 Owners' Workshop Manual
Land Rover 90,110 & Defender Owners' Workshop Manual

Brooklands Books Ltd., PO Box 146, Cobham, Surrey KT11 1LG, England.
Telephone: 01932 865051 Fax: 01932 868803
Brooklands Books Ltd., 1/81 Darley St., PO Box 199, Mona Vale, NSW 2103, Australia.
Telephone: 2 9997 8428 Fax: 2 9979 5799
Car Tech, 11481 Kost Dam Road, North Branch, MN 55056, USA
Phone: 800 551 4754 & 612 583 3471 Fax: 612 583 2023

ISBN 1 85520 2840
Printed and issued in England by
Brooklands Books Ltd.
PO Box 146, Cobham, Surrey, KT11 1LG
England
With the kind permission of
Rover Group Limited
B-LR20HH